Maladies *of* Empire

Maladies *of* Empire

How Colonialism, Slavery, and War
Transformed Medicine

Jim Downs

The Belknap Press of Harvard University Press
Cambridge, Massachusetts
London, England

First Harvard University Press paperback edition, 2023
First printing

Library of Congress Cataloging-in-Publication Data

Names: Downs, Jim, 1973– author.
Title: Maladies of empire : how colonialism, slavery, and war transformed
medicine / Jim Downs.
Description: Cambridge, Massachusetts : The Belknap Press of Harvard
University Press, 2021. | Includes bibliographical references and index.
Identifiers: LCCN 2020018202 | ISBN 9780674971721 (cloth) |
ISBN 9780674293861 (pbk.)
Subjects: LCSH: Epidemiology—History. | Slaves—Health and hygiene. |
Imperialism and science. | War—Medical aspects.
Classification: LCC RA649 .D68 2021 | DDC 614.4—dc23
LC record available at https://lccn.loc.gov/2020018202

For my mother's maternal grandmother, who, a century ago, cut off her hair and sold it so her children could eat, and who passed along to me the power to persist; and for my mother's paternal grandmother, who was born in the nineteenth century with a veil over her face and healed her son's wife by chanting indecipherable prayers over her body and who passed along to me the power to see

AND

For my father's people, who despite subjugation by the British Empire, passed along to me the power to know

AND

For Catherine Clinton, whose unwavering mentorship lighted my way

Contents

Maladies *of* Empire

Introduction

THE SHIP HAD sailed for over a day from the west coast of Africa. All he could hear was the sound of men speaking foreign languages; the water crashing on the hull; the pitiful cries coming from below decks; the wind powering the sails to the Americas. All he could see was the sky.

He had gotten into an argument with a chief, who, in an act of revenge, had accused him of witchcraft and then sold him and his family from Ghana into slavery, to be taken to the New World. He refused to accept his fate. And so, when the members of the ship's crew came along to feed the enslaved Africans with a sticky paste made of beans, rice, oil, and pepper, he did not lift his head to the ladle hovering above him and open his mouth. A crew member noted that he "refused all sustenance." Somehow, he managed to get a knife. And then, in an ultimate act of resistance, he slit his own throat, choosing to die rather than be enslaved in the Americas.[1]

Members of the crew noticed his bloody body and notified the surgeon on board, Thomas Trotter, "who sewed up the wound" and applied a bandage to his neck. That night, the man ripped off the bandage. He yanked at the stitches, trying to wrench them from their hold on his skin. He pulled out all the stitches. Then he dug his fingernails into the

other side of his neck, where he tore a ragged hole in the skin. His fingernails turned red, and he bled profusely.

The next morning the crew discovered him, still alive. They dragged
him up to the deck. Still able to speak, he declared that "*he never would
go with white men*." He then became incoherent and "looked wishfully
at the skies." The crew restrained his hands again and tried to force food
down his throat, but after eight to ten days of refusing to eat, he died.
His name is not known. Other details about his life have not survived.
We do not know if his family members, who were also aboard the ship,
witnessed his act of self-mutilation.

Decades later, in 1839, Robert Dundas Thomson, a physician in
London, recounted the story of the anonymous man dying on the slave
ship in the *Lancet*, a leading British medical journal. Thomson had not
observed this man himself; he was relating a story that had been told
by Trotter during his testimony before Parliament on the slave trade,
in the 1790s. Thomson used this account as one among several examples to determine how long a human being can survive without food.
It wasn't the brutality that mattered to him, though he did recognize it.
It wasn't the violence of the slave trade, though he did detail it. It was
that the enslaved African man had survived for over a week without
eating—that was what mattered to Thomson.

In addition to the story of the enslaved man, Thomson reported other
case studies to illustrate that "insufficient nourishment" caused disease
and to document how long the human body could survive without food,
a condition he called "inedia." Thomson included two other accounts
from the slave trade: he wrote of an enslaved African man who held his
teeth shut when the crew tried to pry open his mouth with a metal in-
strument to feed him, dying after nine days without food, as well as of
a captured African woman who was flogged for refusing to eat and
died after about a week. He concluded that "inhabitants of Africa can
only live without food for ten days" when confined in a crowded, unventilated environment. Case studies that he gathered from Europe
showed that life could be preserved longer under better circumstances;
these examples ranged from a London plasterer to a group of miners
and a pregnant French woman to a mentally disabled Scottish man

(Thomson's own patient) who lived for seventy-one days on water and small beer alone.

✤   ✤   ✤

While it is widely known that physicians drew on patient case studies, it has been less often recognized that the slave trade also provided them with examples. Thomson was able to use examples that developed from the international slave trade to form conclusions about how long humans could live without food under different conditions. The slave trade placed large groups of people in crowded conditions that led to medical disorders and provided doctors with valuable information. It created the extreme circumstances in which people refused to eat, contributing to Thomson's ideas about "inedia."

Thomson's article was published at a time when medical knowledge was undergoing a major revolution.[2] Historical studies traditionally trace these changes to the medical debates that were taking place in London, Paris, and even New York.[3] *Maladies of Empire* shifts the focus away from these cities to the epidemic crises that were unfolding around the world and argues that the development of epidemiology, a branch of medical science that deals with the distribution, spread, and control of disease in populations, developed not just from studies of European urban centers but also from the international slave trade, colonialism, warfare, and the population migrations that followed all of these. While the term "epidemiology" did not become officially recognized until 1850, with the establishment of the Epidemiological Society of London, epidemiological thinking, particularly the creation of various methods to track the cause, spread, and prevention of disease, began much earlier.[4] As *Maladies of Empire* shows, military hospitals and camps, slave ships, and large-scale population movements created crowded conditions that helped physicians visualize the spread of disease and provided different kinds of information from what they learned from observations of cities, prisons, and hospitals.

Understanding the spread of disease took on special urgency as doctors responded to medical crises that erupted from the international slave trade, colonial expansion, and warfare. In treating the populations that were created by these conditions, military and colonial doctors developed theories about the cause, spread, and prevention of disease. The process of centralizing and analyzing medical information about the health of large populations unfolded during the same period that governments in the West were developing mechanisms to wield authority over populations based on new understandings of biology.[5]

Maladies of Empire reveals how slavery, colonialism, and war, often treated separately in scholarly studies, had common features from the vantage point of medical professionals. These episodes produced large captive populations. Slave ships, plantations, and battlefields created social arrangements and built environments that allowed physicians to observe how disease spread and prompted them to investigate the social conditions that led to the outbreak of disease. The increased appearance of these settings around the world between 1756 and 1866 gave way to a proliferation of medical studies that contributed to the emergence of epidemiology. This book begins in 1756 with the story of British soldiers dying in an overcrowded jail cell in India, which served as a touchstone example throughout the medical profession of the need for fresh air. It concludes with efforts made by various countries around the world to track the spread of the 1865–1866 cholera pandemic.

✢   ✢   ✢

In response to medical crises among enslaved Africans, colonized people, soldiers, Muslim pilgrims, and other dispossessed populations, medical personnel developed a set of practices intended to prevent future epidemics. They observed, documented, and named the medical crises. They counted the number of people who became infected and the number who died. They assessed the sanitary conditions and theorized about the cause of the epidemic. They then wrote letters and reports that became part of the burgeoning military and colonial bureaucracy.

While doctors and others had documented health conditions in earlier periods, *Maladies of Empire* explains how ideas developed between 1756 and 1866 became codified into medical theories that contributed to the development of modern epidemiology. It traces how these ideas began first as observations, then as official reports, and finally as arguments and theories in medical journals, lectures, and treatises. Scholars have detailed the ways in which military physicians during wartime innovated sophisticated surgical methods and therapeutics but have overlooked how they created methods that influenced the field of epidemiology.[6]

Military and colonial bureaucracy served a central, often under-appreciated role in advancing the field of epidemiology.[7] Military doctors developed ideas on the ground that eventually evolved into published articles and treatises. British doctors who had worked abroad—from Jamaica to Sierra Leone and from Constantinople to the Cape of Good Hope—were among those who joined the Epidemiological Society of London. When members returned to London to attend meetings, they read their reports about outbreaks of cholera, yellow fever, and other infectious diseases around the world. They also kept an eye on out-breaks of disease in other locations.[8] Building on geographer David Livingstone's contention that the location of scientific knowledge informs the conduct and content of the investigation, I argue that part of the origin story of epidemiology has been overlooked because it re-sulted from studying people who suffered from war, enslavement, and imperialism—most of whom were people of color—in Africa, the Caribbean, India, and the Middle East.[9]

There was a transnational flow of medical knowledge about how dis-ease spread that increased between 1756 and 1866 and transpired not only at familiar hubs of medical research but also at sites of imperialism, slavery, war, and dispossession. Empire, war, and slavery established bu-reaucracies that collected reports on disease, which made it visible.[10] Medical and colonial officials in the British Empire, for example, re-ported, analyzed, and published their findings about the spread of in-fectious disease in the same way that the medical and military officials in the Confederate South and Union North did during the Civil War. The International Sanitary Commission followed a similar pattern of data collection in the mid- to late nineteenth century.

Doctors' reports about global outbreaks of infectious disease provided the medical community with an aerial view of how disease unfolded in a particular region. Military medicine, in particular, established geographic coordinates that enabled disease spread to be mapped. Gaining a panoramic view of an epidemic helped to develop the framework for med-ical surveillance, a key method in contemporary epidemiology.[11]

The expansion of colonialism, particularly in the British West Indies in the mid-nineteenth century, led to a more formalized, uniform method of reporting by military doctors. The outbreak of both the Crimean War and the US Civil War further solidified this practice as military doctors reported on the spread of disease on battlefields and their supervisors collated these documents, interpreted the findings, and gained an overview of a particular region. By the 1860s, epidemiologists had emerged as a defined group of specialists and were able to refine their practices and methods in investigating the 1865–1866 cholera pandemic.

This book further argues that the simultaneous occurrence of the international slave trade, the expansion of colonialism, the Crimean War, the US Civil War, and the travels of Muslim pilgrims had a significant influence on medicine. Until now, these topics have been isolated from each other. *Maladies of Empire* brings them together to explain how they changed the medical profession's understanding of disease transmission. The urgency to study the spread of infectious disease resulted from the confluence of these social transformations. As a result of their work in various parts of the globe, doctors were able to refine theories of contagion that had long been a subject of debate. Slavery, imperialism, and war offered opportunities to study large numbers of people at once. While some of the theories later proved to be inaccurate, their arguments were an integral part of the development of methods—data collection, medical surveillance, and mapping—that remain principal staples of epidemiological practice today.

✦   ✦   ✦

I began work on this book in 2011 as a continuation of my research on the 1865–1866 cholera pandemic, which I examined in my first book, *Sick from Freedom: African American Illness and Suffering during the Civil War and Reconstruction*. I became intrigued by how the medical profession responded to the cholera pandemic, which led me to the National Archives in London. Since then I have followed clues to archives in a number of locations. The copious documents housed in British and US archives were the most useful, but I also gained valuable insights from visiting other sites, even when I did not uncover relevant evidence. In Malta, for example, I was able to see the quarantine facilities that remain standing today, which provided an unparalleled first-hand experience of the structure of the lazaretto on Manoel Island. Each clue led to a new question, often setting the parameters further back in time. The decision to make enslaved people, conscripted soldiers, and subjects of empire the central subjects of the book resulted from the sources I uncovered during my research.

In an effort to excavate the lives of these people from the extant sources, I have relied on Black feminist criticism as my main critical methodology. This method provides useful strategies for how to reconstruct the past by recovering lost subjectivity in the records.[12] Drawing on these interventions has enabled me to reclaim some of the countless people who influenced the advancement of epidemiology. This method was particularly helpful, for example, in my analysis of James McWilliam's investigation of a yellow fever outbreak in the Cape Verde islands in the 1840s (Chapter 3). His report was astounding, as it contained first-person testimonies of mostly colonized and enslaved people describing in detail the symptoms, spread, incubation period, and mortality that resulted from the epidemic. And yet, this rich document required the use of Black feminist theory and criticism to avoid following the internal logic of McWilliam's publication, which cast colonial doctors and British political figures as the protagonists. Drawing on the work of Hazel Carby and Saidiya Hartman, I flipped the script and centered the enslaved and colonized people as the focal point of the chapter. McWilliam's interviews helped to establish the value of collecting patient testimony in order to track the spread of an epidemic; I was able to see this by using these instructive methods.

Colonialism, slavery, and war provided copious evidence about the spread of infectious disease that subsequent generations studied. In fact, case studies like those gathered by Robert Thomson in the pages of the *Lancet* soon disappeared as the origin stories about the development of epidemiology. The theories were remembered, some were codified into scientific principles, and others were discarded, but the predicaments, places, and people who informed their analysis were largely forgotten.

Maladies of Empire is an effort to recover that history and to chart the factors that informed the development of epidemiology. It is also, most importantly, an effort to shift the focus away from medical theorists, doctors, and other professionals to the people whose health, suffering, and even death contributed to the development of medical knowledge—but who have otherwise disappeared from the history of medicine. Their names and voices have often been lost, at times purposely erased from the historical record. *Maladies of Empire* aims to outline the settings that led to their disappearance and reclaim their place in history.

CHAPTER 1

Crowded Places

Slave Ships, Prisons, and Fresh Air

IN JUNE 1756, 146 wounded and fatigued British soldiers were confined to a sweltering prison cell in Calcutta (Kolkata). The cell was tiny and cramped, with only two barred windows. The men were insatiably thirsty; they struggled to breathe and "panted for breath." They stripped off their clothes and fanned themselves with their hats.[1]

Thus begins an account by Robert John Thornton, a British physician, of the tragedy that occurred in what was later known as the Black Hole of Calcutta. Thornton based his account on that of one of the captives, John Zephania Holwell. Holwell was governor of the British garrison of Fort William; when the garrison was captured by the nawab of Bengal, he and his men were thrown into the brig. Thornton's tale focuses on why the conditions proved so deadly.

Some time after they were imprisoned, a guard was convinced to bring water for the desperately thirsty men. The men passed their hats through the bars and the guard filled them with water, but much of the water dripped out as he passed the hats back through the barred windows. Many of the men got no more than a drop of water. They began shouting, "Water, Water!" The stronger men pushed the weaker ones out of their way, killing and trampling some of them.

In the midst of the chaos, Holwell requested to sit alone and die in a corner. His thirst "grew insupportable; his difficulty in breathing increased." He managed to crawl to the window, but then fell to the ground. One of the men noticed that Holwell was still alive; they passed him some precious water, but realizing that he was still thirsty after drinking the water, he decided not to drink any more. His condition slowly improved, writes Thornton, after he was moved closer to one of the windows, "where *fresh air* brought him to life." When the men were finally released, according to Holwell, only 23 of the 146 prisoners were still alive; the rest had suffocated.

Thornton uses this incident to warn of the dangers of crowded spaces and to argue that fresh air is *"absolutely necessary for the continuance of life"* and *"a due supply of it indispensable."*[2]

✦ ✦ ✦

Before the mid-eighteenth century, Western physicians had known that air was crucial to human existence, but they did not understand the details of how it could be compromised in crowded spaces.[3] Early studies on air included those of Robert Boyle, known as the first modern chemist, and Stephen Hales, an English scientist and clergyman who invented a device called the "pneumatic trough," which captured gases emanating from plants. This device allowed eighteenth-century scientists like Joseph Priestley and Antoine Lavoisier to collect and identify the different components of air, including oxygen.

While scientists studied the composition of air in laboratories, physicians began to analyze disease in terms of changes in the quality of air. The British army physician Sir John Pringle, for example, drew attention to the problem of "foul air" in his 1752 book on military diseases. He theorized that many diseases were caused by putrid air emanating from such sources as marshes, excrement, rotting straw, and sick people in hospitals. Pringle also advocated for the use of mechanical ventilators, designed by Hales, to bring fresh air into closed areas.[4]

Physicians also studied the effects of air by turning to people living in confined spaces throughout world. The story of the prisoners of war in the Indian jail appeared in Thornton's book *The Philosophy of Medicine*, published in 1799, as part of a series of examples used to prove the need for fresh air.[5] Ships, like jail cells, noted Thornton, also required the circulation of fresh air. Installing ventilation systems aboard ships, he contended, would help prevent disease outbreaks.

As evidence of the need for fresh air on ships, Thornton cited the observations of Dr. Thomas Trotter aboard the slave ship *Brookes* in 1783–1784. Trotter, a Scottish naval physician, had served in the British Navy during the American War of Independence; after the war ended, he, like many other military surgeons, was redeployed to work on slave ships.[6] As the surgeon aboard the *Brookes,* he not only treated sick crew members but also investigated the physical conditions aboard the ship. Appalled at the cramped conditions for enslaved Africans, he testified before the House of Commons in 1790 during an investigation into the slave trade. A sketch of the enslaved Africans stowed away like cargo at the bottom of the *Brookes* became one of the most iconic, if problematic, images of abolition throughout the Atlantic World.[7]

While the diagram of the *Brookes* continues to be included in exhibits on slavery and abolition to illustrate the brutality of the slave trade, few people remember that Trotter's observations on the *Brookes* helped solidify early scientific understandings of the importance of oxygen to survive.[8] Trotter's discoveries grew out of his investigation of the architecture of the ship, a scurvy outbreak, and the rise of pneumatic chemistry.

When the *Brookes* arrived on the Gold Coast of West Africa in June 1783, the crew purchased "more than a hundred prime slaves, young stout and healthy." Members of the crew shackled the enslaved Africans in pairs, tying chains around their arms and legs. They forced them onboard and into the bottom of the ship, where there were no windows or ventilators, and the air had already grown stale. The ship stayed on the coast for a month while more enslaved Africans were acquired; it was bound for the Caribbean and the southern United States, where the Africans would be sold.[9]

Before the ship set off across the Atlantic, the crew remarked that the Africans were "growing exceedingly fat." To strengthen them for the trip, they were being fed beans, rice, and corn, boiled together with palm

DESCRIPTION OF A SLAVE SHIP.

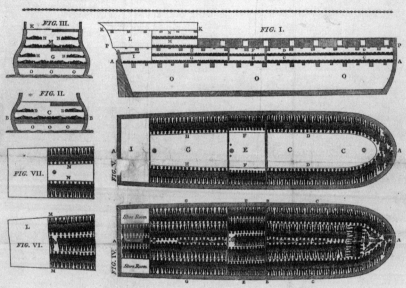

The interior of the *Brookes,* showing how the enslaved Africans were packed tightly into the hold. (British Library, 522.f.23)

oil, Guinea pepper, and salt, and served as a paste. The enslaved Africans had some access to water, but they were confined to the bottom of the ship "sixteen hours of the twenty-four" and not permitted any exercise. The "custom of dancing them around the deck to the sound of the drum," wrote Trotter, "was not practiced till it was too late to be of service."[10]

Trotter predicted that the Africans would not remain healthy under such conditions. They were housed in rooms five to six feet high, "imperfectly aired by gratings above." The temperature reached above ninety-six degrees Fahrenheit. Some were "stowed spoonways . . . closely locked into one another's arms" on "the sides of the vessel, raised about two feet and a half from the floor, and of breadth sufficient for the length of a man." When he went below deck, Trotter remarked that it was "difficult to move without treading upon them" and that he struggled to breathe. No precautions were taken "to preserve the health of the slaves."[11]

At one point, an African man whom Trotter described as "corpulent" complained of a hardness on his right arm. By the next day, the hardness had spread to the upper part of his forearm, and he experienced "some contraction at the joint of the elbow." Eventually it spread to the shoulder, neck, and lower jaw. Once the enslaved man's neck was affected, he became delirious. His tongue was "locked between the teeth" and then lolled out of his mouth for three days before he died.[12]

Trotter tried to treat him. He placed him in a warm bath and forced his jaw open to administer medicine, but it was ineffective. When Trotter inspected his mouth, he noticed that his gums were "spongy" and "bleeding," that his back teeth were loose, and his breath was "extremely disagreeable."[13] Once he noticed these symptoms, he was convinced that the man was suffering from scurvy.

As the ship continued its voyage across the Atlantic Ocean, others began to present similar symptoms. Trotter noticed that some of them were the "fattest slaves" aboard the ship and postulated that being overweight increased susceptibility to scurvy.[14]

While it is unclear how Trotter communicated with the enslaved Africans, they helped to shape his understanding of scurvy.[15] He learned from them to identify the early stages of the infection: the hardness in the arm, the immobility of the elbow, the tightening of the jaw, how the pain increased as the disease progressed. He saw that the limbs weakened and the desire to sleep became strong, progressing to "coma and delirium."[16] Trotter's analysis of scurvy depended primarily on his observations of these enslaved Africans, which added essential detail to what he had observed among British sailors. The enslaved Africans were not simply passive objects of Trotter's study but rather conveyed to him, either through a translator or nonverbally, how scurvy affected their bodies.

By April 1784, three months after the ship had departed from Africa, seven to eight people had died. Trotter reported that the number of enslaved Africans with scurvy continued to escalate: "Some were found dead in the rooms in a morning or dropped down immediately on coming upon deck, while others expired eating their victuals, full in flesh and blood." As the *Brookes* approached the Caribbean, forty more deaths occurred. Of the roughly six hundred enslaved Africans who arrived in the Caribbean, Trotter diagnosed three hundred of them as having some degree of scurvy.[17]

Trotter noticed that the Africans, like other scorbutic patients he had observed, displayed a strong desire for acidic foods, and during the voyage, he decided to do an experiment to see if he could confirm the effectiveness of their preferences: "Having repeatedly observed the scorbutic slaves throw away the ripe guavas, while they devoured the green ones with much earnestness, I resolved to try if there were any difference to be remarked in their effects." He divided a group of nine Africans with symptoms of scurvy into groups of three and gave limes to one group, green guavas to the second, and ripe guavas to the third. After a week, those who had eaten the ripe guavas had not improved, "while the others were almost well."[18]

Once the ship arrived in Antigua, and then when it stopped in St. John, Trotter procured fresh vegetables and fruits to feed those who had scurvy. Sucking the juices directly from oranges, lemons, and grapefruits was the "surest way for securing virtues of the citric acid," he claimed.[19] The enslaved Africans were also unshackled, since they were no longer considered to pose a threat to the crew.[20]

Once the Africans began to eat citrus fruit, their symptoms quickly dissipated and their health improved. By the time they reached Jamaica, he noted, "there were little remains of Scurvy among them: they were now better fed, and repaired for market." Trotter was convinced that if they had not been given fruit in Antigua, at least half of the enslaved population would have died within ten days.[21] As the surgeon aboard the ship, he had fulfilled his mission: he had prevented a scurvy epidemic from jeopardizing the profit that investors hoped to make from selling enslaved Africans to work in plantations across the Atlantic World.[22]

✦ ✦ ✦

Although hired to serve as the surgeon on the *Brookes* to protect enslaved Africans' health and to ensure that they were ready "for market," Trotter used his experience to develop new ideas about scurvy. When he returned to Edinburgh, he shared his observations with one of his professors, William Cullen, who lectured on scurvy at the university.

"Hearing at the time so much said on the subject of Scurvy, a disease that I had lately treated in a multitude of cases," wrote Trotter in the introduction to his book, "I began to think that there were many valuable facts in my possession that might go a great way in deciding the dispute."[23] His observations of scurvy on a slave ship gave him information that was different from what could be obtained from naval ships because of the extreme conditions under which the Africans were held. Enslaved Africans were confined in the bottom of unventilated ships, fed poorly, and not allowed to exercise. These conditions, Trotter concluded, contributed to the severe outbreak of scurvy.

In 1786, after returning from his voyage, Trotter published *Observations on Scurvy* in London; a revised edition appeared in 1792, and the book was also printed in Philadelphia and translated into German. In the book, Trotter discussed existing views on the causes of scurvy and proposed his own theories. Decades earlier, in the 1740s, the Royal Navy surgeon James Lind had demonstrated that consuming oranges and lemons could cure scurvy. Lind is often regarded as the person who discovered the cure for the disease, but as Trotter explained in the introduction to his treatise, the question had not been settled; there remained many different opinions about the causes of scurvy and how to treat it. Trotter not only used his observations of enslaved Africans to corroborate Lind's theory, but disputed alternative theories and discussed factors that increased susceptibility to the disease.

Trotter argued against the belief that meat, alcohol, and opium cured scurvy and addressed a number of other misconceptions, including theories that scurvy was a disorder of the blood and that it was a contagious disease. He did not attribute the prevalence of the disease among the Africans to differences between the races. Scurvy, he believed, resulted from poor nutrition, not racial identity; the white crew had access to fresh vegetables, while the Africans did not. His only reference to racial difference was to note that the "livid coloured spots" characteristic of the disease were not noticeable on black skin.[24]

Trotter confirmed Lind's theories about the curative power of citrus fruit, but he contended that Lind's method of preparing the juice was ineffective. The best method was to have patients suck the juice directly from the fruit; fresh juice could also be preserved if prepared properly.[25]

Trotter identified a number of contributing factors that made the enslaved Africans particularly susceptible to scurvy, such as a poor diet and lack of exercise. In addition, "the predisposition to Scurvy must be also increased by hardships they experience from their confinement in the hold of a ship."[26] Long before medical anthropologists coined the term "structural violence" to describe how poverty, violence, and other forms of oppression increase the incidence and severity of disease, Trotter recognized that the pain and anguish of enslavement could predispose people to become sick.[27] He focused on the slave trade itself rather than on individual characteristics or racial identity. The brutality and violence of being transported from Africa to the Caribbean was devastating: "It would be unjust to suppose that the African feels no parting pang when he takes the last farewell of his country, his liberty, his friends," he explained. A number of women had "violent hysteric fits" during the brutal voyage, and "hideous kinds of moaning" could be heard from below decks at night. Some of the Africans committed suicide by jumping off the ship rather than be sold into slavery in the Caribbean.[28]

Trotter emphasized that the brutality and violence of the international slave trade made the African people more susceptible to scurvy. He believed that the epidemic was exacerbated by an unhealthy diet, insufficient water, and lack of exercise, combined with the poor conditions of living in a crowded space in the bottom of the slave ship. He thus advocated for fewer Africans to be placed on future vessels and for higher quality food, citrus fruit, and adequate water to be supplied. He also urged slavers to incorporate more exercise regimens, even suggesting "they may likewise be encouraged to dance, which, besides the exercise it affords, tends to amuse them; diverts their ideas, and makes them more cheerful." But, he warned, these recommendations should be executed gently and not by "the lash of a cat-and-ninetails."[29]

Trotter's focus on the horrific conditions that enslaved Africans endured certainly inspired his analysis, but he was also influenced by the development of pneumatic chemistry in the eighteenth century, which led many in the medical community to consider how crowded environments changed the quality of air.[30] He encouraged the use of larger ships to transport fewer enslaved Africans in order to ensure the circulation of air, and he stressed that grain in the ship's storerooms should be kept as "airy as possible."[31]

Trotter's study of scurvy both relied on and contributed to studies in pneumatic chemistry. "The present area of chemistry," he wrote, "has unfolded a treasure of knowledge." He theorized that acidic fruits were effective in treating scurvy because they contained "vital air," or "oxygene." He also emphasized the importance of fresh air, noting that the "tainted atmosphere of the slave rooms," which was full of "foul air" and "impure exhalations," predisposed the enslaved Africans to illness.[32]

Trotter's analysis depended upon his observations of enslaved Africans on the *Brookes,* but when he published his treatise he referred to them in the introduction as "a multitude of cases."[33] This phrase legitimates his study by highlighting its reliance on empirical observations, but it obscures that many of these "cases" were actually enslaved Africans. Writing for an audience of his medical professional peers, he uses clinical language that erases how the conditions at the bottom of the slave ship created an unhealthy environment that enabled him to develop an argument about the cause, cure, and prevention of scurvy. As a result, enslaved Africans disappear as key players in the creation of new scientific knowledge.

Further, the crowded conditions at the bottom of the ship helped to make the need for fresh air visible and served as scientific proof for the emerging study of oxygen. While it may seem like common sense to conclude that the crowded conditions and a lack of fresh air could lead to death and disease, it was not an accepted fact at the time.[34] In *The Philosophy of Medicine,* Robert Thornton cited Trotter's description of conditions on the slave ship as one of several examples to prove the necessity of "a supply of air." He then mentioned the case of another slave ship on which, according to a Mr. Wilson, 586 of the 2,064 enslaved Africans whom his ship had purchased to be brought to the Americas had died during the journey, presumably from overcrowding.[35] Thornton also used the case of the British prisoners in Calcutta to make this point, which further shows that prisons and slave ships, along with hospitals and jails, helped to make the medical community aware of the need for ventilation.[36] Trotter's evidence about conditions aboard a slave ship informed an important debate among medical thinkers and legal authorities.

In 1790, Trotter appeared in front of a select committee of the House of Commons to testify against the international slave trade. He reported on the lack of air below decks on the *Brookes*, affirming that he could never breathe there. "I have often observed the slaves drawing their breath with all the laborious and anxious efforts for life, which are observed in expiring animals, subjected by experiment to foul air, or in the exhausted receiver of an air-pump," he stated. He described them "crying out in their own language, 'We are suffocated.'"[37] His testimony served both to support the abolitionist movement and to advance the understanding of the need for fresh air.[38] He ended his treatise on scurvy by laying out rules for nutrition, exercise, and cleaning of enslaved people's quarters on ships, and concluded, "To all these rules it is necessary to add, that the most lenient means should be employed to keep them in order; great address is often wanted on such occasions; but severities are substituted, and few think that they are trampling on the Rights of a Man and a Brother!"[39]

Trotter's medical analysis of both scurvy and the effects of the lack of air at the bottom of the ship provides an important case study in the development of an epidemiological approach to disease. He made recommendations to political authorities on the prevention of disease, which would later become routine in epidemiological practice. Further, he identified social conditions that contributed to sickness and suffering. He also studied the health of a population rather than an individual, a foundation of public health. The phenomenon of enslaved people becoming sick on ships also showed how man-made environments and human decisions could lead to the spread of disease.

Trotter's testimony, in which he quoted enslaved Africans describing in their own language that they were suffocating, appeared in medical treatises on the importance of oxygen. In the 1796 edition of *Medical Extracts*, Robert Thornton, who was a proponent of pneumatic medicine, devoted part 2 of the first volume to "The Agency of Oxygen Air in the Animal Body, and the Cause of Vital and Voluntary Action." He discussed Trotter among other well-known scientists and doctors, like Priestley and Lavoisier. Their studies were based on laboratory experiments, while Trotter's resulted from his observations on a slave ship; but in late eighteenth-century medical discourse, both appeared side by side in medical studies. The subtitle of Thornton's book, *On the Nature of Health, with Practical Observations: and The Laws of the Nervous and Fibrous Systems,* which included sections on diet, exercise, and clothing, among other topics, highlights how the value of Trotter's observations served as significant evidence for the medical community to understand the "agency of oxygen," which had developed as a new field of study due to the rise of pneumatic chemistry. "Practical observations" often referred to how physicians drew on particular patient medical histories as case studies to support their arguments. While historians and other scholars have remembered Priestley and Lavoisier for their contributions to the study of oxygen, they have overlooked how Trotter's study validated their findings by offering evidence based on his observations of enslaved people.[40]

✛ ✛ ✛

Interest in the importance of fresh air in slave ships coincided with the more well-known rise of prison reform efforts that began in the late eighteenth century with John Howard, a British philanthropist and prison reformer. As high sheriff of the village of Bedford, England, Howard was tasked with inspecting the county jail. He quickly discovered that the prison charged inmates a fee for being incarcerated but that many were unable to pay, which only prolonged their sentence. He found this practice appalling, particularly in the case of inmates who might have been innocent or who were imprisoned for minor offenses.[41] Shocked by this and other conditions in the jail, he spent the winter of 1773–1774 inspecting prisons all over Britain and, later, on the European continent.

During his tour, he followed a systematic routine of collecting and recording his observations. After a day in the jail or prison, he would spend the night in a local inn and document the conditions that he had observed. He counted how many rooms were in each facility, measured their size in feet and inches, and recorded how many men shared a room. He noted which prisons allowed men to interact with women during the day, which jails included intellectually disabled people among the general population, which ones had young boys.

In the process of documenting these social and physical conditions, he also recorded the health conditions. Many jails, he found, did not supply water. Even in those that did have water, inmates had to rely on the jail keeper or servants to provide it to them. And even then, they were only given three pints a day for both drinking and cleaning. Howard then took note of the air. Providence, he wrote, provides air to all people for free. It requires no labor for inmates to receive it. Yet prisoners are robbed of this "genuine cordial of life." Even animals need fresh air to survive. "Air which has performed its office in the lungs," however, is "feculent and noxious." In discussing the importance of fresh air, Howard referred to the case of the British prisoners who suffocated in the Indian jail in 1756.[42] His use of this example indicates how this case served as a touchstone for people at the time who were thinking about the need for fresh air, and how examples from other parts of the world informed the development of medical knowledge in the United Kingdom.

Howard believed the lack of basic necessities—clean water, fresh air, and suitable living conditions—led directly to the outbreak of infectious disease. "Gaol fever," as well as smallpox and other illnesses, he claimed, resulted from the crowded conditions and poor architecture.[43] Gaol fever, or jail fever, was an eighteenth-century medical classification for fevers that broke out in jails and other confined places. While it may have described a number of different diseases, it is generally considered to refer to what we now call typhus.

Like many people at the time, Howard believed that infected inmates spread disease to others.[44] He quoted from a treatise by James Lind, the surgeon who had introduced lemons as a cure for scurvy, claiming that "the source of infection to our armies and fleets are undoubtedly the jails," and that British ships that were sent to fight in the American War of Independence lost two thousand troops because of disease introduced by men who had been impressed out of prisons.[45]

Based on his observations of prison conditions, Howard made recommendations for improvement. He argued that inmates should be placed in individual cells to prevent the ill effects of impure air, which overcrowded jails produced.[46] He underscored the significance of "fresh and sweet air" and "open windows" and advised that prisoners be "made to go out and air themselves at proper times."[47] Howard's re-

port ultimately persuaded government officials to redesign prison lay-
outs. In 1779, the House of Commons passed the Penitentiary Act,
which called for cities and boroughs to construct prisons providing
single cells for inmates based on Howard's ideas. When local magis-
trates built new prisons, they drew on Howard's recommendations of
individual cells.[48]

Howard continued his investigation of prisons and traveled throughout
Europe, inspecting prisons in Holland, Germany, Russia, and elsewhere. In a prison in Moscow, he witnessed men chained to the walls
but noted that the women did not suffer similar abuses. He visited a
military prison in Butyrki, where prisoners were crowded into a single
room and had "pale sickly countenances." He also toured the military hospital founded by Peter the Great, which included a prison. He
described the ward as "dirty and offensive" during his first visit, but
when he returned later with a Russian physician, who escorted him, it
was "much cleaner."[49]

In England, Howard's reforms led local and state officials to routinely
inspect prisons in an effort to improve health conditions. These reform
efforts continued to reveal that crowded conditions led to the spread
of disease. Jails helped to make disease outbreaks visible; they illustrated
the leading medical theories about the spread of infections and the necessity of fresh air.[50] This knowledge, in addition, contributed to a more
sophisticated understanding of how infectious diseases spread, which
subsequently helped to lead the way to sanitary reform. Even though
we now consider early theories about the cause of jail fever to be wrong,
these theories were important for later knowledge production.[51]

Once physicians understood jail fever as a result of crowded conditions, they were able to develop the mechanisms to cure it. When officials framed the disease as a natural, predictable outcome of prison life,
it thrived. Framing jail fever as a medical disorder that resulted from
crowded conditions ultimately led to its decline within prisons.[52] A
century after the publication of Howard's treatise on prisons, an article
in the *Lancet* by R. M. Gover, medical inspector of prisons, traced the
connections between crowded environments and the spread of disease.[53] Published in 1895, at the height of the so-called bacteriological
revolution, when germ theory had revolutionized ideas about the cause

of disease, the article focused on Howard's prison reform efforts as crucial, pioneering work in the prevention of jail fever.[54]

The *Lancet* article also commended Thomas Trotter's work on "jail fever" aboard ships. Gover recounts an outbreak of fever that Trotter described on the HMS *Colossus* in 1796, which had a crew composed of many men who had recently been released from jail. Gover quotes a British physician who praised Trotter for showing that fevers very similar to those in jails will break out on ships "when unwholesome conditions prevail, as starvation and cold, a scanty supply of impure water, deprivation and pollution of air, exclusion of light, idleness, filth, listlessness and grief."[55]

Reformers like Trotter and Howard were not alone. Scores of other physicians stationed throughout the world provided key information about the dangers of crowded conditions and the need for fresh air. Noted British reformer Edwin Chadwick, for example, often celebrated for his efforts in London, had a global perspective. In his 1842 report for the government on sanitary conditions in Great Britain, he noted that impoverished conditions in Constantinople as well as "filth" and "closeness" in Paris caused high mortality rates.[56] Physicians and reformers working in England and France communicated with each other, and John Howard toured prisons in several European countries. After his work on prisons, Howard turned to conditions within hospitals. He visited hospitals throughout Europe in 1785–1786, including the Hôtel-Dieu de Paris, the leading hospital in France. He described the patients' rooms as "nasty and offensive." Two patients were often forced to share a single bed, no windows were open, and "medical gentlemen were prejudiced against a free circulation of air."[57]

Following Howard's critique, Jacques-René Tenon, a French surgeon, wrote a report in 1788 condemning the unsanitary conditions of the hospital, drawing specific attention to the lack of ventilation. "We have seen rooms so narrow that the air stagnates and is not renewed," he wrote, "and that light enters only feebly and charged with vapors." Sick patients were housed in the same rooms with those who were dying, and those suffering from noncontagious disease were often exposed to those who were contagious. Patients walked barefoot to a nearby bridge for "fresh air" in the winter months.[58]

The poor condition of the hospitals propelled French revolutionaries to establish a new system of medical care. Drawing on ideas of rationality and order that grew out of the Enlightenment, they initiated a number of changes, including segregating the sick from the indigent poor, the dead from the living, and the men from the women, which Howard had observed in French prisons.[59] Their efforts led to the construction of more hospitals throughout France that followed guidelines promoting proper ventilation and hygiene.

Additional work in late eighteenth-century France highlighted the need for sanitary reform and fresh air in prisons and hospitals, based partly on Lavoisier's work on oxygen.[60] François-Joseph-Victor Broussais, a leading French physician, became interested in the cause of disease, particularly the ways in which oxygen could overstimulate the body's organs. Cold air, food, drugs, or vapors, he argued, could lead to overstimulation in the organs and cause irritation that led to fever.[61] Like many other physicians who devised new ways of thinking about medicine and therapeutics, Broussais had spent time in the military. He spent three years as a naval surgeon in the war against England and then became a physician in Napoleon's armies, serving for over a decade in Spain, Germany, Holland, and Italy. During this period, he developed theories about disease causation.[62] His understanding of climate in foreign areas informed his later understanding of tuberculosis. In a discussion of the distinctive lesions produced by this disease, he wrote, "If we now seek for the circumstances which prepare and facilitate the formation of these tubercles, we will be satisfied, 1st, that they are common in cold and moist countries, and rare in warm climates, even in constitutions which are attacked with them in cold regions. This I have verified, as I have repeatedly said, during twenty years of military practice."[63]

When a cholera epidemic broke out in 1832 and spread throughout the world from India to Europe to the Americas, Broussais's status as a leading figure in the field suddenly declined. His theories, which emphasized physiology and inflammation of the organs, could neither explain the cause of the epidemic nor offer a theory on how to treat cholera.[64]

Meanwhile, the French physician Nicolas Chervin, after studying yellow fever in the United States, the Caribbean, and South America, returned to France in 1822 as a leading authority on the subject. Based on his time abroad, he believed that yellow fever was not contagious but was caused by a miasma, a vapor that could transmit disease through the air but not by direct human contact.[65] A few years later, when cholera broke out in France, Chervin remained committed to his anticontagionist argument, which asserted that cholera did not spread from person to person, despite explosive spread of the disease and a rising death toll. Chervin died defending his ideas but left an important imprint on the field of growing field of epidemiology. Although Chervin did not discover the waterborne origins of cholera, his efforts to consider the environment and the local circumstances contributed to later epidemiological research on the disease.[66]

French physicians in the early nineteenth century, although they did not succeed in understanding the underlying causes of cholera and other diseases, made advances in statistics, physiology, and anatomy.[67] Their scientific understandings were limited and, at times, inaccurate but their investigative approaches were innovative and influential. They succeeded in creating protocols to maintain sanitary hospitals and to promote healthier environments. Local governments established health councils, and citizens created voluntary organizations to promote healthier environments. Their work led to actual changes in the landscape of the country: dead animal bodies were removed from city streets and country roads, stricter policies were developed regarding sewers and drainage, and the number of public baths was increased. In addition to pinpointing sites that they believed produced disease, the physicians also, more importantly, initiated a public discourse that alerted the citizens and the government to the problems of poor health and cast blame on those who they believed contributed to the rise of these conditions.[68]

French physicians created ways of talking about disease transmission that facilitated the emergence of epidemiological principles.[69] The hygiene movement identified health dangers of crowded spaces. Their discussions of crowded spaces provided the foundation for the development of epidemiology by offering a new way of thinking about the air. It moved beyond the ancient understanding of miasma, which people throughout Europe and other parts of the world believed caused epidemics to spread, and instead emphasized crowded spaces as the cause of disease outbreaks. While some proponents of miasma developed ideas about "night air" and "foul air," they may have, at times, pointed to crowded spaces as the cause of disease but their focus remained more tightly attentive to the air's movement that emanated from rotten vegetation or corpses not simply by the physical environment. French and British physicians moved away from an analysis focusing purely on air as the central factor in the cause of disease and instead emphasized how physical conditions of crowded spaces caused disease to spread.

Many British and French doctors advanced the study of the spread of infectious disease among subjugated people in other parts of the world. Colin Chisholm, a British surgeon who served in the War of American Independence and then in Grenada, from 1783 to 1794, discussed the dangers of crowded conditions in a book he first published in 1795 (with a second edition in 1801) on a yellow fever epidemic in the West Indies, in which he argued that the disease was contagious. According to Chisholm, the "lower class of the inhabitants of St. George's, Grenada," especially "sailors, soldiers, and sailor and porter negroes," engaged in illegally selling and drinking rum in small buildings and lanes. There, "they soon become intoxicated, and are crowded together in a hot, putrid, or infected atmosphere, till they recover their senses; when they generally find themselves precipitated into a fever of a most malignant character."[70]

The value of Chisholm's book lies less in its scientific validity than in how it put forward a theory that drew on first-person observations of crowded conditions. Chisholm's observations of the "lower class," consisting of impoverished white sailors and Black porters, led him to understand how crowded and cramped conditions encouraged the spread of disease. The living conditions within the Caribbean served as evidence for Chisholm's argument.[71]

As a number of scholars have shown, many physicians at the time viewed poor people as innately susceptible to illness and as responsible for spreading disease.[72] Impoverished populations, however, became the subjects of medical study. The importance of fresh air and ventilation, which became one of the main tenets of public health, emerged from observations of poor people's crowded living conditions. Poor people, including enslaved Africans and prisoners, made disease visible and became the subjects that physicians turned to in order to understand how epidemics developed. Because Grenada was part of the British Empire, Chisholm had access to the intimate living arrangements of the white and Black people he described.

Just as Howard outlined measures to prevent disease in jails, Chisholm proposed a number of preventive measures. One set of measures involved improving living conditions and sanitation in Grenada. He called for destroying the small wooden homes and rebuilding them in stone and brick. Narrow and dirty streets should be replaced by spacious ones that would allow air to circulate. Butchers should slaughter animals far from the center of towns to keep "putrid offals and meat" from spreading disease, and their sheds and stalls should be well-ventilated and have access to running water. Cemeteries should also be established far from the center of town.[73]

✛   ✛   ✛

The expansion of the British Empire to other parts of the world, particularly India, led to the proliferation of more case studies like that of Chisholm, which contributed to the development of an epidemiological approach to disease. As more British physicians sailed to India in the nineteenth century, they began to study the effects of the tropical environment on health.[74] While much of the subcontinent remained outside the purview of British military physicians, they did have ready access to jails and prisons. In these built environments, they systematically classified diseases and symptoms and tabulated illness and mortality rates, engaging in the kind of statistical thinking that informed the development of epidemiology.[75] Even though Trotter, Howard, and others published widely known studies, not every physician was familiar with their ideas about how crowded spaces led to the outbreak of disease or knew the case of the British prisoners of war suffocating in the Indian jail. While on colonial missions, many military doctors created theories in response to the crises they confronted.

The rise of colonialism—like the expansion of the international slave trade—led physicians to produce medical ideas outside of the metropole. In an 1852 article on the health of British troops in India, for example, Major David Grierson, a British military physician who was inspector general of hospitals for a region of India, wrote about the issue of ventilation and fresh air in barracks and hospitals.[76]

Grierson discussed a report he had previously written about the jail in Karachi, where he noted that "the space allowed for patients was very small, and seemed to have been really productive of very bad effects." He proceeded to explain how much space should be allotted to each individual in jails, hospitals, and barracks, to be certain they have enough fresh air. In determining the minimum amount of space a person needs, he quoted the British physician David Reid as recommending 1500 cubic feet but noted that others in the medical community argued that that amount of space was "an extravagance." Grierson remarked that in Europe, men were given an allowance of at least 1000 to 1500 cubic feet, with good ventilation, but in warmer climates, like India, the space ought to be even larger.[77]

Grierson acknowledged that some officials disagreed with his conclusion, on the grounds that the allowance was well under 1000 cubic feet per person in hospitals in London, Dublin, and Edinburgh, as well as in the provincial hospitals "in all of England and in the General Hospitals, Naval and Military, of the British Government."[78] His critics, he wrote, also seemed to imply that Europeans required more space than "Natives." Grierson disagreed, based on arguments by Lavoisier and a variety of early nineteenth-century scientists who made calculations about the relationship between space and the availability of fresh air.

Grierson referred to other case studies to support his argument that air quality could be compromised by factors such as "humid air, high temperature, the privation of oxygen in respired air, as well as carbonic acid gas." To show that water vapor derived from exhaled air is noxious, he quoted a report stating that in "the houses of the poorer classes in Russia, where the windows are single, and a number of persons occupy a small stove-heated room, a thick icy crust forms on the inside of the windows during frosty weather, arising from the condensation of the breath, perspiration, and the aqueous fumes of candles, and of the stove, &c. When a thaw comes on, the icy crust is converted into water and a deleterious principle is disengaged, which produces effects similar to those arising from fumes of charcoal." The affected people are carried out of the house to the open air, rubbed with snow, and forced to drink cold water until their "natural colour is restored."[79] Grierson contrasted this with situations where "purer air" was available. Grierson's focus on the "houses of the poorer classes" further underscores how the Russian poor, like the British poor, in previous examples, helped to illustrate a theory about the changing quality of air.

Grierson discussed various formulas for designing buildings to ensure proper ventilation, concluding that the space allowed for one individual should be 3000 cubic feet with natural ventilation only, or 1800 cubic feet with strong mechanical ventilation.[80] Alongside this quantitative analysis, Grierson also used the story of the impoverished Russians to argue for the importance of fresh air, allowing readers of the journal to visualize this process. The story of the Russians reveals a phenomenon that could only be partially explained by a mathematical equation.

Military medical officials in colonial regimes, like Grierson, turned to specific case studies to illustrate the cause and spread of disease. They also increasingly turned to statistics and engineering principles in order to show the danger of crowded spaces. Like later generations of epidemiologists in the twentieth century, these physicians devoted as much time and effort to understanding the spread of disease as they did in trying to help prevent it.

Histories of disease prevention and sanitation often focus on reform efforts to clean up gritty urban centers from London to Paris to New York, where people were crammed into crowded tenement buildings, where pigs roamed among the impoverished, where excrement clogged city streets.[81] Municipal reform campaigns did popularize theories about how crowded conditions prevented the circulation of fresh air and promoted disease, but these were not the only sites that caused concern among physicians and sanitarians.[82]

Colonialism and slavery also produced crowded conditions that gave rise to concerns that predated mid to late nineteenth-century urban reforms, often considered the major turning point in sanitary reform. Colonial expansion placed doctors in new locations where they devised medical theories about the cause and spread of infectious disease. The rise of the international slave trade, which involved confining large numbers of human beings in crowded spaces, produced medical crises that then helped to give rise to new medical theories. Although slave ships were a crucial site of investigation, they often entered medical journals and reports, like Trotter's treatise on slavery, simply as "cases" or as "ships," which had the unintended effect of erasing slavery from the discussions of the importance of fresh air for health.

Lauded for his device that captured gases emanating from plants, Stephen Hales invented ventilators to be used on ships but often did not specify that many of the ships carried enslaved Africans. In 1741 he presented his first paper to the Royal Society illustrating the importance of these devices, which promoted fresh air. In the paper, which is over two hundred pages long, he made only passing references to slavery. In order to explain the number of men required to operate the ventilators, he stated, "suppose there be in a Transport, or *Guinea* Slave-Ship, two hundred Men, as there is often about that Number," then each man, he claimed, would only need to operate the ventilator for a half hour every two days.[83]

In a second paper delivered to the Royal Society over a decade later, in 1755—a year before the infamous incident of the Black Hole of Calcutta—Hales more explicitly referred to slave ships as key evidence to prove the value of ventilators on ships. In this paper, titled "An Account of the Great Benefit of Ventilators in Many Instances, in Preserving the Health and Lives of People, in Slave and Other Transport Ships," he provided several remarks from authorities on many ships, including slave ships and ships transporting convicts and settlers, who testified to the importance of ventilators. One captain informed him that his ship had departed from Guinea, West Africa, for Buenos Aires, Argentina, with 392 enslaved people, and that all of them, except twelve enslaved Africans who entered the ship sick with "the flux," which was likely dysentery, survived the deadly trip across the Atlantic Ocean.[84]

Hales argued for the use of ventilators for ships, which were designed, as he explained in his subsequent 1758 treatise, to "preserve the Health and Lives of Men in Ships; as also the Ships from decaying, when laid up in Ordinary." "And as the Malady in both Cases," he asserted, "principally arises from damp, foul, stagnant, close confined putrid Air, so the obvious and only Remedy is, to exchange the Air of Ships so often, as not to give it Time to putrify."[85] Ventilators promoted the circulation of the air and were thought to prevent the spread of "jail fever" in crowded ships, hospitals, and prisons.[86]

In order to provide evidence of the need for ventilators, Hales drew on examples from slave ships. "I was informed that in a *Liverpool* Ship, which had Ventilators, not one of the 800 slaves died," he noted, "except only a Child, born in the Voyage." He further added that "in several other Slave-ships, without Ventilators, there died 30, 40, 50, or 60 in a Ship."[87] In the 1758 treatise Hales also included a letter from a French doctor who claimed that ventilators on ships decreased the mortality rate from "one-fourth of those valuable Cargoes, in long Passages from *Africa* to the *French* Plantations," to less than "a twentieth."[88] While it was not uncommon for physicians to include letters from medical officials as supporting evidence to buttress their argument, this particular letter reveals how the medical study of crowded conditions on slave ships extended beyond the British Empire to the French Empire.

The presence of enslaved people on the ships provided Hales with compelling evidence of the need for ventilators. In his 1755 paper, he included a letter from a captain of a slave ship who lamented how few ventilators were used on ships that carried "Passengers, Slaves, Cattle, and other perishable Commodities." The captain remarked that once he began using ventilators, he buried "only six" slaves during a fifteen-month voyage. The captain further submitted that the remaining "340 Negroes were very sensible of the Benefits of a constant Ventilation, and were always displeased when it was omitted."[89]

The captain's inclusion of the testimony of enslaved Africans provides a subtle but powerful clue on how enslaved people contributed to the development of medicine and science. By 1756, medical professionals had begun to realize that crowded conditions caused the spread of disease. The accounts from British prisoners of war in India provided first-hand testimonies of the dangers of crowded spaces. The reports in Hales's publications provided observations by those on slave ships who had witnessed how crowded conditions led to inordinate mortality. According to Hales, his ventilators solved the crises aboard all ships, not just slave ships. Ventilators promoted circulation that safeguarded the transportation of passengers, goods, and crops, but enslaved people's demands for the crew to run the ventilators provided powerful evidence of the efficacy of ventilators on these ships.

Compared with passengers, sailors, and soldiers, enslaved Africans faced the most extreme and violent conditions on ships; therefore, their cries for fresh air within this medical context provided compelling evidence about the efficacy of the ventilators.[90] In general, ship authorities and the crew ignored the crying pleas of enslaved Africans and watched many die onboard the ships. In this case, the captain did not necessarily care about the enslaved Africans on the ship but simply reported their pleas to illustrate that the ventilator provided relief. Enslaved Africans' cries for fresh air got translated into Hales's medical report as an endorsement.

Hales's analysis of the conditions of enslaved Africans during the international slave trade, nonetheless, was of central importance for his 1758 book *A Treatise on Ventilators: Wherein an Account Is Given of the Happy Effects of Many Trials That Have Been Made of Them.* In it, he reprinted the evidence from the earlier reports that included examples from slave ships. But read in isolation, one would not necessarily know how fundamental slavery was to his thinking, since the case studies are somewhat buried in an otherwise highly arcane medical treatise. Read in isolation, one would not know that this treatise was a more refined theoretical sequel to an analysis grounded in the material reality of slavery. One would not know that enslaved Africans' cries became proof that compelled officials to consider ventilators. One would not know that the international slave trade led to a major new advancement in the construction of naval fleets.

The elision of slavery from the title of Hales's treatise could be simply the result of popular rhetorical practice: there was no need to distinguish slave ships from others in the royal fleet. But when placed alongside Trotter's treatise on scurvy, the absence of slavery seems less accidental or even customary and more a result of a pattern: as the scientific ideas evolved from first-hand accounts to medical theories, slavery seems to fall into the category of "cases," in Trotter's book, or "trials" in Hales's case, as in the subtitle of his treatise.

These cases reveal how slavery led medical thinkers to focus on the health of large groups in confined conditions.[91] While outbreaks of epidemics had in the past also led doctors to explore the occurrence of disease in terms of a larger public, the increased prevalence of ships, jails, and plantations provided more opportunities to study the dangers of crowded spaces and contributed to the development of public health by framing health in terms of the collective body. British reformer Edwin Chadwick may well have drawn on evidence from slavery and colonialism when writing his authoritative report on sanitation, even though he included no references to these forms of subjugation. He quoted a physician from the London Fever Hospital: "In dirty and neglected ships, in damp, crowded, and filthy gaols, in the crowded wards of ill-ventilated hospitals filled with persons labouring under malignant surgical diseases or bad forms of fever, an atmosphere is generated which cannot be breathed long, even by the most healthy and robust, without producing highly dangerous fever."[92]

Many of the eighteenth- and early nineteenth-century case studies on crowding and fresh air came out of military reports from British physicians based in colonies in the Caribbean and India, or on slave and naval ships, but when others took up these conclusions, they often ignored the context under which they had been produced. Medical professionals knew that many transatlantic voyages involved ships carrying enslaved Africans. Further, they knew that hospitals and prisons did not simply refer to those in the British metropole but also those in the Caribbean and India, which informed medical analyses such as Grierson's reports. Yet, when Chadwick wrote his report on sanitation, he did not refer to these case studies.

Physicians and medical reformers not only investigated how disease spread among populations in crowded spaces but also began to turn to people around the world to test theories about contagion and infection. Yet, as we will see, similar to the history of crowded spaces, fresh air and ventilation, the people—the washerwomen and hospital workers—who helped doctors visualize these cases are missing from traditional accounts of the history of medicine.

CHAPTER 2

Missing Persons

*The Decline of Contagion Theory
and the Rise of Epidemiology*

W E DON'T KNOW THE NAME of the laundress who washed the
dirty linen of travelers quarantined on the Mediterranean island
of Malta in the 1830s—travelers who had arrived by ship from plague-
ridden cities like Alexandria.

We do know that she lived in Valletta, the capital of British-controlled
Malta, probably near the gardens that overlooked the harbor. Passengers on ships from North Africa and the Middle East that stopped in
Malta, a hub for maritime traffic in the Mediterranean, were required
to undergo quarantine before continuing on to Europe. Those on ships
that either had known cases of disease or were suspected to be carriers
were given a "foul bill of health" and quarantined for a longer period.
While some crew members remained on the ship, most passengers disembarked and were lodged in the lazaretto, a large stone hospital and
quarantine facility on a small island in the main harbor of Malta.[1]

Soiled linen from the ship would be collected and delivered to the
laundress, who was quarantined along with the passengers in the lazaretto. Working in the dank basement, she would likely have hauled
buckets of water and armfuls of wood to boil clothing in large vats. She

would soak the clothes overnight to begin the process of removing the stains and smell. The next morning, she would begin the arduous process of scrubbing the fabric. She would use alcohol to remove bloodstains, chalk and pipe clay to remove grease stains, and even urine as bleach, grinding these potpourri ingredients into the clothing.[2] The clothing would then be hung outside to dry in the Mediterranean sun.

Laundresses in Malta followed this routine of cleaning soiled linen for decades. The clerk of the lazaretto in the 1830s, Giovanni Garcin, noted their routines, their labor, and their health. Despite being in direct contact with the dirty linen, he testified, none of the laundresses he had observed in his twenty-nine years of employment had become infected with plague. This testimony appeared in a treatise by the British physician Arthur T. Holroyd that presented an argument against the practice of quarantine. The fact that plague did not spread from soiled linen to laundresses was one piece of evidence Holroyd brought forward to prove that plague was not contagious.[3] The treatise was addressed to Sir John Cam Hobhouse, a British member of Parliament who spearheaded various reform efforts and was president of the Board of Control for India. Holroyd contended that existing quarantine regulations were unnecessary, expensive, and outdated.

The laundresses likely had no idea that they served as an example to help understand how disease spread (or didn't spread). Who the laundresses were, what they thought as they entered quarantine, and whether they feared working with linens from infected ships did not matter to those who observed them. The reference to the laundresses in Holroyd's treatise appears on just two pages, but it was an important clue in undermining the prevailing belief that plague was contagious.

Such bits of evidence were not uncommon throughout the late eighteenth- and early nineteenth-century quarantine debates. They appear as short statements, sparsely worded phrases, parenthetical notes, and elliptical references in reports by doctors and quarantine officials about infectious diseases. These people—laundresses, Muslim pilgrims, sailors, and the poor—helped the medical community and government agencies visualize how disease spread. The authorities did not need to know the names of the dispossessed or ask them how they understood disease.[4]

✦   ✦   ✦

The medical community depended on populations throughout the empire to understand infectious disease. In the eighteenth and early nineteenth centuries, the quarantine debates provided a crucial forum for physicians to rethink widely held medical beliefs. Writing in journals and letters to colleagues, medical authorities often relied on single episodes, anecdotes, and narrative descriptions as case studies. Europe provided many examples of disease spread and quarantine, but colonialism in Asia, the Caribbean, and the Middle East further increased the range of examples and contributed to the development of particular methods of gathering information.[5] Plague no longer existed in England, but the expansion of the British Empire enabled physicians to study outbreaks that occurred elsewhere. The movement of ships, people, and cargo provided intellectually curious doctors with huge sample sets to investigate disease outbreaks. In addition to ships, they turned to colonial hospitals and other settings. Physicians also called on local doctors, healers, and colonial officials for data. The bureaucracy of colonialism provided a formal system of documenting and collecting information that flowed directly to government authorities. A number of physicians wrote treatises that aimed to repeal the legal requirements for quarantine.[6] By studying the spread of disease throughout the world, mapping its coordinates, pinpointing its origin, and defining its behavior, physicians developed key epidemiological methods.

During this period, many physicians began to doubt the theory that underlay quarantine regulations. Plague, cholera, and certain other epidemic diseases had for long been thought to spread through contagion—direct contact with a sick person or with objects the person had come into contact with. Battle lines were drawn up between contagionists and anticontagionists, the latter supported by merchants who wished to do away with the costly quarantine system.[7] To disprove the contagion theory, physicians often turned to hospital workers, who, like laundresses, had close proximity to the sick. The hospital workers' health became a barometer to determine the contagiousness of disease.

One physician who took careful note of hospital workers was William Twining, a British military doctor who was assistant surgeon at the Calcutta General Hospital from 1830 until his death in 1835.[8] He published an influential, comprehensive volume on diseases in India in 1832. The chapter in his book on cholera discusses the disease's progression and symptoms, cause, and treatment, as well as whether or not it is contagious.[9] To investigate the question of contagiousness, Twining made use of his experience in the hospital, detailing instances of people who came into close contact with cholera patients. He turned first to hospital attendants and those who handled the linens: "The persons most exposed to contract Cholera in the General Hospital at Calcutta, (if the disease were contagious)," he wrote, "are those having charge of the bedding and clothing, and those employed in personal attendance on the patients." He went on to list, by name, the men who changed and washed soiled bedding. They included the current "clothes keeper," Shaik Selim; Selim's predecessor, Dhowall, who had worked in the hospital for twenty-three years; the head washerman, Gawhee; and his predecessors, Hassye and Beechuck, the latter of whom had been a washerman for twenty-one years. "Not one of the subordinate washermen or people employed about the clothing and bedding stores has ever had Cholera," Twining reported.[10]

Twining then discussed several other groups of people who had close contact with cholera patients in the hospital. "Native dressers," including the head dresser, Buctourie, who had worked at the hospital for twenty-six years, changed bandages applied after patients were bled or treated with leeches; "sweepers" cleaned bedpans "as well as the pans in which the matter vomited is received"; "Hindoo coolies" put the blankets back on patients who threw them off, rubbed their limbs, and could not avoid inhaling their breath "in the most deplorable stages of the disease." Finally, a number of Indian medical students had cared for patients when the hospital was crowded and the attendants overwhelmed. No one in any of these groups, claimed Twining, had ever been attacked by cholera.[11]

An unusual feature of Twining's book is that he named many of the Indian hospital workers: Shaik Selim, Dhowall, Gawhee, Hassye, Beechuck, and Buctourie. These men served as unofficial, and more than likely unasked, participants in Twining's analysis of how cholera behaved. The daily labors of the hospital's most subordinate class offered important evidence that cholera was not transmitted through direct contact. The names provided a human shape to a scientific postulate and enabled the reader to imagine the proximity the washermen and attendants had to infectious disease.

Twining's book was one of a number of works in which medical professionals used dispossessed, often colonized peoples throughout the world to help visualize the spread of infectious disease. Many of these physicians became part of a growing faction of anticontagionists.[12] One of these was Amariah Brigham, an American physician who published a treatise on cholera in 1832 and discussed the evidence for and against the contagion theory.[13] He anthologized information from physicians around the world in order to buttress his claim that cholera was not transmitted by direct contact either with the sick or with contaminated material. Presenting evidence from doctors in India, Russia, Poland, Prussia, England, France, Canada, and the United States, he focused on people who had close contact with cholera patients, especially hospital attendants. Although admitting that attendants occasionally became ill, he contended that they were no more liable to contract the disease than were other laborers. He cited a Dr. Searle, for example, who worked in a hospital with many cholera patients in Warsaw and reported that "not ONE of the hospital attendants—not ONE of the nurses—not ONE of those who handled the dead, fell victim to the disease." Similarly, eight Russian physicians remarked that "the attendants who nursed and applied frictions to the patients—who put them into baths, changed their linen, and performed other offices for the sick, remained free from cholera." While some of the physicians Brigham cited mentioned other groups that did not become infected despite close contact with cholera patients—soldiers, nursing babies, physicians, household members, and fellow patients—hospital attendants were the most frequently mentioned.[14]

✦   ✦   ✦

Efforts to observe the health of hospital attendants and launderers as a way to understand how disease spread were prevalent in the early nineteenth century, as physicians began to work around the world and write reports about their experiences. Physicians also mentioned the ethnicity and nationality of people in locations where epidemics started, as well as of those who appeared to be more susceptible to disease. Amariah Brigham collected reports from various colonies to track the spread of the 1819 cholera epidemic from India to Southeast Asia and Africa. Describing the outbreak among enslaved people on the island of Mauritius, he wrote, drawing from an article by the physician Charles Telfair, a British official and plantation owner stationed on Mauritius, "The patients consisted principally of the black population of Mozambique caste. These are curly headed Negroes; low in the scale of civilization; too unintelligent to learn trades, and generally employed as porters."[15] Brigham identified the causes of cholera as "impure air, low and damp dwellings, crowded and filthy houses and cities, poor diet, intemperance, fear, &c. &c."[16] Throughout Brigham's analysis of cholera, which he also traced to North America and Europe, he argued that it resulted from dirty, crowded conditions and "damp surroundings."[17] The case of enslaved people proved an analogue to impoverished people living under similar circumstances in Europe and North America. Brigham's treatise reveals how enslaved people served as key "patients" to illustrate his theory about how and why disease spread. Medical authorities examined outbreaks not only in European and North American cities but also among those living under slavery in various parts of the British Empire.

In Holroyd's book on plague and quarantine, people of color are specifically referred to in a few instances as evidence to support his anticontagion theory. In a series of questions addressed to Thomas Leslie Gregson, chief surgeon of the naval hospital in Alexandria, Holroyd asked if Gregson knew of any examples of people who had contact with a plague patient but did not contract the disease. Gregson replied that he knew of many hospital attendants who had not become sick, and several people who died of the plague and were visited by friends who were not attacked. He was then asked if he knew of any cases of a hospital attendant who had contracted plague but had not communicated it to patients. Yes, he replied: "Amongst others a servant belonging to our hospital was attacked; he was black. We were shut up in Quarantine, one thousand in number. Other three of our servants were attacked, the last on the seventh day of our being shut up. These four had their huts together, and had been infected at the same time. Whoever has seen an Arab hut could easily conceive this happening, even in Typhus." Two of the four died, but no patients in the hospital were infected.[18]

Another person of color who served as evidence for Holroyd was mentioned in the questionnaire answered by Henry Abbott, a British physician and colleague of Thomas Gregson in Alexandria, who provided an example of a person who came down with the disease without having had contact with any other plague victims. In 1835, Abbott reported, he had been on a gunship that had been quarantined for six weeks with no plague cases. The first person attacked by the plague was "a black," a prisoner from Nablus who had been taken on board at Jaffa.[19] Another physician questioned by Holroyd, Dr. Pruner, who thought that plague could sometimes be transmitted by contagion, explained his theory of how the 1835 plague epidemic had started in Cairo: "The Plague of 1835 was imported in the first instance, by the brother of Ciglio, a Maltese Physician—transported by him to another brother— by the second brother to a black woman—from her to a Greek neighbour &c."[20]

Throughout this period, references to people of color regularly appeared in medical journals and scholarly articles without reference to racial difference or inferiority but rather as markers to map the spread of disease. When Henry Dickson, an American physician based in South Carolina, wrote a lengthy treatise on pathology and therapeutics, he highlighted the case of a Black patient in order to show that dengue, a disease later identified as a mosquito-borne viral infection, was contagious. He cited the work of a prominent physician who claimed that "a negro" was the first patient in the city to contract dengue, having acquired it from the captain of a Cuban ship that had anchored in Charleston. Dickson also discussed the spread of dengue from St. Thomas to St. Croix. He cited a Dr. Stedman, who reported that the first patient brought the disease to the island and spread it within his family; it then spread from family to family and estate to estate, "exactly in proportion to their contiguity, or to the intercourse that might happen to exist." To demonstrate this pattern, he noted that the disease had spread from enslaved people on one estate to those on another that was several miles distant but had the same owner—presumably because the enslaved people moved from one estate to the other.[21] While the focus on spread among enslaved people could connote an argument about racial difference, Dickson did not make such claims.[22] In describing the pathology of dengue, for example, he stated that all patients have a similar "exanthem," or widespread rash: "All classes of person were subject to this singular exanthem, and all equally and alike. The aged and the young, the infirm and the robust, the native and the stranger, the black and the white, all shared the same sufferings."[23]

Physicians often used Black and other dispossessed populations to illustrate medical theories about the spread of disease. In his study of pathology and "the origin and laws of epidemics," a Maryland doctor, Moses L. Knapp, argued that different groups of people were more or less well adapted to different climates. "Transfer a New Orleans negro to Canada," he wrote, "and he falls a victim to the climate in a year or two; the same of the Canadian who takes up his abode permanently in New Orleans."[24] Unlike many writers of this period, he does not make the argument based on racial inferiority.[25] The "New Orleans negro" serves more as a figure to illustrate a medical theory about climate than an argument about racial ideology. The use of the term, nonetheless, points to Knapp's assumption that his readers would recognize Black people in New Orleans as the embodiment of race. At the time of the publication of the book, New Orleans served as the major hub of the domestic slave trade and contained a large population of enslaved Black people. By conjuring the image of a "New Orleans negro" and not just any enslaved or Black person, Knapp draws on the cultural iconography of slavery in order to illustrate his theory about climate.[26]

٭ ٭ ٭

The early nineteenth-century physicians who studied the spread of disease by observing poor and colonized people built on their predecessors' theories about the dangers of crowded conditions in slave ships, prisons, and hospitals; though their arguments were similar, they reached their conclusions through the creation of nascent epidemiological methods. Let us return to Arthur Holroyd, who argued against quarantines on the basis that plague was not contagious. Holroyd investigated the topic by creating a questionnaire for doctors and administrators who worked in British hospitals and quarantine centers in Egypt and Malta. Questions concerned everything from sanitary conditions to incubation time and contagion.

Holroyd posed questions to Thomas Gregson, the British physician stationed in Alexandria who had reported that hospital attendants did not become infected with plague. When asked if he had observed "disease propagated by contagion," Gregson stated that he had not seen any cases and that in fact, "on investigation have found many so reported to be false."[27] Gregson's use of the term "investigation" is an indication of the shift to a more scientific approach to medicine that was taking place in the nineteenth century. Doctors began to see themselves as investigators who carefully explored the physical and natural world to pinpoint the cause of disease and understand how it spread.[28] Gregson's investigative spirit even led him to study a disease outbreak in animals. Summoned to examine an illness that had killed over one hundred of the pasha's oxen, he examined the animals, discovered gangrene and buboes, declared that they had died of the plague, and "caused them to be interred deeply."[29] In collecting information on how plague spread, Holroyd also asked the colonial physicians whether they knew of any cases where the disease had been "communicated by sexual intercourse." Pruner, Gregson, and Abbott all replied no, with the latter two mentioning cases they knew of where a plague victim had not infected a sexual partner. Holroyd cited another case, which the captain of the Malta lazaretto had told him about, of a man who was not attacked by the plague even though he "had connexion" with his wife shortly before she died of the disease.[30]

The questionnaire's format enabled Gregson and the other doctors to document their insights, observations, and analyses. In these narrative spaces, doctors followed well-established patterns of using dispossessed people to track the spread of epidemics. In the case Gregson reported about the "black" servant at his hospital who came down with plague, he argued that the infection resulted not from contagion but from the servant's living conditions. Three other servants became infected at the same time, and "these four had their huts together."[31] Gregson's comment that anyone who has seen an Arab hut "could easily conceive this happening" illustrates how familiar images of the Middle East were among physicians, and how such places could be easily conjured in their collective imagination.[32] Commenting on Gregson's observation, Holroyd concluded, "Egypt is never entirely free from plague, partly from atmospheric causes, and partly from local circumstances dependent upon the physical condition of the people . . . ; but mainly to the small, close, confined and huddled huts which compose an Arab village, and which are always the hot-beds of the malady whether sporadic or epidemic."[33] After the epidemic ended in Cairo, according to Dr. Pruner's response to Holroyd's questionnaire, the local government cleaned the main streets and prohibited burials in the town, but they did not "attend to the ventilation of the houses . . . distribute food to the poor [or] . . . cleanse the houses of the poor."[34] Colonialism in Egypt allowed Gregson and other European physicians to witness and describe the conditions that they believed fostered the spread of plague.

As part of his concluding remarks, Holroyd faulted the Board of Health of Alexandria for inefficiency, "inasmuch as it has never collected evidence respecting the plague, to show how the disease might be diminished or eradicated by improving local circumstances and the physical conditions of the people."[35] It is notable that Holroyd did not fall back on a racial argument that blamed the spread of disease on the Arab population, but instead blamed it on those with power. He also did not put forth a climatic argument. Instead, he pinpointed the failures of the city policy and the poor structure of housing as the cause. The poor people's living conditions in the Arab huts helped to make this claim visible. Holroyd and others argued against contagion theory by contending that plague resulted from unsanitary conditions and crowded spaces.

Being abroad encouraged some physicians to question medical dogma.[36] Because of colonialism, doctors like Gregson had more ability to investigate the link between social conditions and health. Holroyd's treatise on quarantine built on a growing body of medical literature by doctors who were stationed in India and other parts of the British Empire and were focused on the physical and social environment, including the climate and conditions of dirt and overcrowding, as the cause of illness.[37] By focusing his investigation on what doctors and administrators in colonial settings had discovered about the health of communities, rather than individuals, Holroyd was operating like a public health investigator.

Holroyd's major objective was to present a convincing argument against quarantine restrictions, which a growing number of physicians had begun to view as ineffective, by showing that plague was not contagious. We have seen the evidence he presented that the disease had not been communicated to laundresses in Malta or hospital attendants in Egypt. To further illustrate this claim, he turned to Muslim pilgrims returning from the hajj, the annual pilgrimage to Mecca. Pilgrims returning from Mecca on ships to various Mediterranean ports were quarantined for up to several weeks in the lazaretto in Malta. Holroyd quoted Captain Bonavia, superintendent of the lazaretto, who described what happened on one ship full of pilgrims that stopped in Malta in February 1837. The ship arrived from Tripoli, Lebanon, with twenty-one Muslim passengers and a crew of eleven men. Because Tripoli was experiencing a plague outbreak, the ship was kept isolated in the harbor. Two pilgrims became ill with plague after arriving in Malta. One died on the ship, and the other was escorted to the lazaretto with two healthy companions; he died soon thereafter. The rest of the pilgrims were then disembarked and kept in quarantine for forty-one days, while the crew remained on the ship. One crew member died of plague ten days after the pilgrims had disembarked; the remaining crew were then taken off the ship and quarantined for another two weeks. None of the remaining pilgrims and crew became ill.[38]

Holroyd concluded from this report that the plague could not be contagious, as it was not communicated from the plague victims to the others, despite their close quarters. The crew member could not have contracted the disease from the two stricken pilgrims, since he did not become sick until ten days after having been in contact with them. A more reasonable conclusion for why the crew member became infected, wrote Holroyd, was that "the vessel had not been properly ventilated" and "an impure atmosphere still existed." The ship was sunk and then raised up again; when the crew returned, they remained healthy. "Surely it was the removal from the contaminated air, and the purification of the vessel which prevented the extension of the pestilence."[39]

Holroyd argued that both people and objects contaminated with plague could be "purified" if exposed to fresh air. In response to his questions, Captain Bonavia stated that no guardians who handled baggage or merchandise had ever come down with the plague. This included people who handled cotton, but any cotton articles coming from ships that had plague on board were "exposed to the air and ventilation before they are handled."[40]

Holroyd also reported data, obtained from Captain Bonavia, on the number of "passengers, troops, and pilgrims" who were quarantined in Malta. Between 1832 and 1837, Bonavia reported, roughly 10,000 passengers, 3,000 troops, and 2,000 pilgrims had been housed in the lazaretto. For the period from 1810 to 1832, the lazaretto clerk, Giovanni Garcin, reported that an average of 800 to 1,000 people were quarantined annually. No one in the lazaretto, reported Bonavia and Garcin, had ever contracted the plague while in quarantine (unless coming from an infected ship).[41]

The lazaretto officials' documentation of the number of people who were placed in the quarantine facility in Malta is an example of the kind of recordkeeping that became an important element of epidemiology. The simple task of recordkeeping, tallying the number of people in the facility, provided empirical evidence about how plague spread. The counting of Muslim pilgrims, in particular, reflects a larger pattern of physicians depending upon populations throughout the empire to think through medical conundrums. The annual migrations of the pilgrims, which had been documented by both the British and the Russians, often raised fears concerning the spread of disease from one location to another.[42] Their religious journey, in turn, provided information for those studying epidemic disease. Yet Holroyd's discussion of the pilgrims, like his discussion of the Maltese laundresses and Twining's of Indian hospital workers, was less concerned with their individual health conditions than with the evidence their experience provided about the contagion theory. Holroyd, Twining, and other anticontagionists were trying to develop a more rational understanding of the spread of disease, which undermined the argument about contagion.

With the benefit of hindsight, it is possible to say that Holroyd was both right and wrong in his belief that plague is not contagious. While fleas, not humans, are primarily responsible for transmitting plague, it can be transmitted from person to person through the air in its pneumonic form. Holroyd's contribution lies less in the accuracy of his argument about transmission than in the methods he used.

Holroyd condemned Egyptian officials for not collecting evidence about plague. He believed that if physicians had recorded their observations, municipal officials would be better equipped to respond to the epidemic: "Surely medical men, who have seen much of the plague, ought to have been summoned to give all the information they possessed about the malady, with a view, if possible, to lessen the present oppressive and tyrannical laws." In response to Holroyd's questionnaire, Robert Thurburn, British consul in Alexandria, had stated that "the collection of evidence respecting the plague has never, so far as I know, been one of the objects of the Board of Health."[43] Similarly, Captain Bonavia reported that the Board of Health in Malta did not collect evidence from "medical men" about incubation period and other facts about the plague. The board, he stated, "acts according to the general regulations established and observed here from old date, without references to other Medical evidence, excepting that of the local Physicians when occasion requires it."[44]

Holroyd's focus on using "evidence" to determine quarantine laws reflects the growing role of physicians as investigators. His report, which included questionnaires and interviews with officials in Egypt, Malta, and Greece, emphasized that physicians and others on the frontlines of the plague outbreak were best positioned to provide the proof needed to revise quarantine regulations. By questioning physicians and others on site, gathering evidence from a large region, and analyzing the cases, Holroyd was able to draw conclusions about the factors that fostered the spread of plague and to suggest preventive measures.

The value of his report lies in helping to illuminate an overlooked aspect in the history of the British Empire's approach to the study of infectious disease. Holroyd's work follows a well-established pattern among British physicians in the 1830s to 1860s. Deployed across the empire, many were assigned to provide medical care to troops and others in the colonies, and as they responded to medical crises in unfamiliar environments, they became investigators. The debates surrounding quarantine laws in the British metropole and elsewhere during this period provided further incentive to document health conditions, collect evidence, and write reports. In the process of writing these reports, British physicians often observed the spread of disease in populations throughout the empire, like the laundresses in Malta, the Indian hospital attendants, the inhabitants of the Arab villages, and Muslim pilgrims. Their efforts, in turn, contributed to the development of methods that would come to be standard in modern epidemiology.

✢ ✢ ✢

In 1846, Gavin Milroy, a Scottish physician, published a short book on the subject of plague and quarantine. Milroy had served as a medical officer in the Mediterranean and the West Indies, and in 1850 would become one of the founding members of the Epidemiological Society of London. The book comprises a summary of an extensive report that a committee of French physicians had submitted to the Académie Royale de Médecine, preceded by an introduction by Milroy.[45]

Like Holroyd, Milroy believed that quarantine regulations were based on the outdated idea of contagionism and that the time had come to overhaul them: "The absurdly foolish and most ridiculous principles which [the Quarantine rules] embody, the vexatious and oppressive restrictions which they impose, the wretchedness and suffering which they almost necessarily give rise to, and the great increase to mortality which, we have reason to believe, they often occasion, are surely sufficient grounds for the scrutinizing investigation that is so generally demanded."[46] Anticontagionists such as Milroy needed to provide evidence that plague did not spread simply by human contact. The movement of Muslim pilgrims, thousands of whom traveled to and from Mecca every year, unwittingly offered an ideal case study.

Milroy's summary of the French report includes quotes from an article by the French physician Louis Aubert-Roche, who spent a number of years working in Alexandria. Aubert-Roche noted that every year, groups of Muslim pilgrims from "Morocco, Darfour, Egypt, Constantinople, Persia, Asia Minor, and Syria converge at Djeddah, at Medina, then at Mecca, the central point. They carry merchandize with them, for this pilgrimage is also a fair." Despite this gathering of people from a wide variety of places, carrying with them many items, "from time immemorial, the plague has never been seen in Arabia," even in 1825 and 1835, when epidemic plagues broke out in Lower Egypt.[47] Like Holroyd, this report avoids an argument that relies entirely on religious, racial, or ethnic identity to explain the spread of disease, which many at the time believed, although it does refer to "the very peculiar predisposition of negroes to contract the plague."[48] The example of the pilgrims, along with a number of others, was used to conclude that plague was not transmitted either by direct contact or through materials.

In his introduction to the report, Milroy discussed the difference between infectious and contagious diseases. The term "contagious," he wrote, should be used for diseases like syphilis, ringworm, and rabies, which are transmitted only when a person comes into direct contact with the "diseased part in the sick person, or matter taken from it." Such diseases "are incapable of contaminating the atmosphere." Milroy defined "infectious," by contrast, as referring to diseases that are propagated by a "particular effluvium or miasm" that emanates from a sick person's body and is then taken into the lungs or stomach of another person. Infectious diseases included whooping cough and scarlet fever, which "propagate themselves by infecting the atmosphere." Milroy also identified a third category, including diseases like measles and smallpox, which he called "contagio-infectious"—these could be transmitted through both contagion and infection.[49]

By collecting reports about disease outbreaks around the Mediterranean, Milroy and other British and French doctors solidified the distinction between contagious and infectious disease. Their work ultimately contributed to policymakers' decisions to revise quarantine laws. In discussing the nature of infectious diseases, Milroy added an observation that, he claimed, had been overlooked by most writers on the subject: "Whenever a number of human beings, even in a state of health, are cooped together in a narrow ill-ventilated space, the air gradually becomes so contaminated by the effluvia given off from their bodies that . . . fever will almost inevitably make its appearance." He gave as an example the well-known case of the British prisoners of war who died in

Calcutta almost a century before and then added, "We have daily illustrations of the same fact in what takes place on board troop and slave ships, in jails, crowded penitentiaries, and so forth."[50]

This simple statement distills details presented in numerous military and medical reports. For Milroy and his readers, slave ships and jails both provided evidence about the cause of disease outbreaks. Like his eighteenth-century predecessors, he drew on reports of people throughout the empire to make his argument.

∗ ∗ ∗

Throughout the archival record on contagion and quarantine laws, physicians appear as leading subjects, but evocations of an unnamed laundress, slave ships, Indian hospital attendants, and Muslim pilgrims were pivotal in prompting medical authorities to rethink long-held beliefs about contagion. It is easy to miss references to these people in lengthy treatises, and their role has been overlooked by historians. Military bureaucracy was essential for developing the tools that would become standard epidemiological practices—interviewing local medical and administrative officials, collecting reports across a broad geography, establishing modes of surveillance, and creating ways to observe and even map the spread of disease.

Medical writers drew on these examples in order to give human shape to their arguments. Because the people whom they referred to had little social capital or economic clout, they were easily forgotten. Lost to time, they disappeared from the page almost as quickly as they appeared. They have largely vanished from the scientific debates that would later build on their presence or dispute it. Later theories that would dispel contagion theory or consider the quarantine debates as byzantine would forget the people who informed these debates.

While many physicians stationed throughout the world depended upon the movement and health of people throughout the empire to inform theories about contagion, the use of these people also contributed to the development of epidemiological practices. As we will see in the next chapter, an epidemic outbreak among enslaved and colonized people in Cape Verde in 1844 would inform a major global debate about quarantine, the spread of infectious disease, and, most of all, the methods required to ensure the public health of a community. Washerwomen would again emerge as leading actors in this historical episode. While their voices have, until now, been mostly muted or muddled, at best, the voices of the enslaved and colonized, of the washerwomen and soldiers, appear more clearly in Cape Verde. Their testimonies mark a turning point in how subjugated people informed the emergence of epidemiological methods.

CHAPTER 3

Epidemiology's Voice

Tracing Fever in Cape Verde

IN LATE 1845, an epidemic exploded on Boa Vista, one of the Cape Verde islands, off the coast of West Africa. The cause of the epidemic was debated by the enslaved and free population on the island, as well as by the Portuguese, who governed it, and the British, whose vessel, the *Eclair,* had made a stop at the island on its way home to England from Africa. The British government had dispatched the *Eclair* to police the coast of West Africa, watching for any sign of the illegal slave trade. Even though slavery had been abolished throughout the British Empire in 1833, the practice of buying Africans and transporting them to the Americas continued through the 1840s and beyond.[1]

Before the *Eclair* arrived in Boa Vista on its return trip to England, the island had been without any sign of disease. After the *Eclair* departed, however, a few people began to develop an array of symptoms. The sunlight that shone so brightly into their rooms began to bother their eyes. Their muscles ached, but not from the typical labors that their bodies performed—hauling laundry, harvesting crops, hoisting bags of coal, or standing guard at the military fort. Walking, or even standing upright, was laborious. They had no appetite, and even the smell of food made them nauseated. They developed a fever and head-ache, and their bodies convulsed. Some became delirious. Members of their family noticed the peculiar color of the vomit that flecked their lips. It was black.

For some, the symptoms intensified, and in no more than a week, they died. Others recovered without knowing what prayers conjured over their body or which therapeutics—bleeding, enemas and purges, qui-nine, diaphoretics and antispasmodics—cured them.[2] Slowly the mys-terious illness began to spread across the island. Soon, many of the resi-dents had a family member or knew of someone who died of it. British and Portuguese officials in Cape Verde estimated that of the 4,395 resi-dents of Boa Vista, 311 died. Up to two-thirds of the entire population became infected.[3]

When the epidemic first broke out in Boa Vista, there was no public health agency to collect information on where the fever began and how it spread. Instead, there were rumors—fragments of evidence that resi-dents began to piece together to develop a story: a dozen sick sailors on a British vessel; a washerwoman's daughter dead after suffering from the black vomit; a Black military guard attacked by the fever after helping to carry two corpses for burial.

✦   ✦   ✦

The *Eclair* left England in November 1844 and reached West Africa in January 1845.[4] For several months the ship cruised up and down the coast, deploying small boats to patrol the rivers for any trace of the il-legal slave trade. The mission was unsuccessful, which meant no prize money for the crew. Some of the men slept on the boats or on shore at times; a few of them became sick with diarrhea, others with fever, and some died, but this did not alarm the attending physicians initially. In general, naval physicians, along with the public, realized that many crew members who traveled oceanic voyages to unknown lands would be-come sick and die. Only half of most crews were expected to survive.

The ship proceeded to Sierra Leone in July, during the rainy season, where the crew were put to work cleaning out another British ship, the *Albert,* which had been part of the disastrous Niger Expedition of 1841–1842. A number of the dispirited men went ashore on leave; a handful deserted. The *Eclair* eventually departed from Sierra Leone at the end of July. Then the fever broke out in earnest. Many men became ill, especially those who had been on shore, and seven died. They were diagnosed as having contracted "remittent fever," a common disease of tropical climates.

With fever now raging on board, the steamship loaded some coal at Goree, a hub of the international slave trade in Senegal, but French authorities refused to allow it to dock. The *Eclair* then steamed to the Cape Verde archipelago.

When the *Eclair* arrived in Boa Vista on August 21, 1845, the captain notified the Portuguese authorities that he had fever aboard and feared he might be subject to quarantine. A local British surgeon, Dr. Kenny, came on board to inspect the ship. Kenny diagnosed "the common coast fever," which was not thought to be contagious, and Portuguese authorities agreed to allow the ship to dock. Crew members were first kept on the ship, but as more became ill, they were removed to a dilapidated fort on a small island in the harbor, where the healthy and sick were quartered separately. The officers were lodged in the main town, Porto Sal Rey, and some of the sailors were given leave to visit town. Several dozen local laborers cleaned and whitewashed the ship and restocked it with coal and water. The British consul collected a dozen bags of the officers' dirty linen, which he distributed to seventeen local washerwomen.

Meanwhile, the fever continued to spread among the crew of the *Eclair*. Nearly sixty more cases appeared and two dozen men died, including the assistant surgeon. The captain himself fell ill. Three weeks after the ship had docked at Boa Vista, on September 13, 1845, the decision was made to return to England. Crew members, both sick and healthy, left the fort and reboarded the ship, and the *Eclair* departed for England via Madeira. During the return trip, the captain and twelve more crew members died.

When the *Eclair* arrived in England on September 28, 1845, the vessel was quarantined for a time, while more deaths occurred onboard. Debate raged about the imposition of the quarantine. The press, medical professionals, and naval authorities all argued about what should be done, with opinions based on the commercial repercussions of quarantines, the contagiousness of the disease, and the responsibility to the beleaguered sailors. Finally, the surviving crew members—only a third of the original number who had set out nearly a year before—were allowed to disembark.[5]

Meanwhile, in Boa Vista, an epidemic of fever was sweeping the island. The local people believed it had been brought by the *Eclair* and were demanding compensation from Britain, but the Portuguese government was reluctant to jeopardize relations with its ally and declined to assign blame. The British decided to pursue an investigation anyway, to put to rest any suspicions that might lead to stricter quarantines against British ships.[6] Sir William Burnett, director general of the Naval Medical Service, tapped the navy surgeon James Ormiston McWilliam to lead the inquiry. McWilliam would travel to Cape Verde to explore all aspects of the outbreak.

To prepare his report, McWilliam interviewed over one hundred people on Boa Vista, almost all of them people of color, including nearly a dozen enslaved people. The population of the Cape Verde islands at the time consisted chiefly of Afro-Portuguese people—descendants of enslaved Africans and Portuguese settlers. McWilliam documented the population of Boa Vista as consisting of 434 enslaved people, 3,875 free native people, 84 Europeans, and 2 Americans.[7] The testimonies, which McWilliam transcribed, mark the most extensive surviving record from the nineteenth century of people of African descent describing in detail the onslaught of an epidemic in the Atlantic world.[8] Those who fell ill related when they became sick and who they had been in contact with, and described their symptoms. Others detailed the experiences of their neighbors and relatives.

McWilliam's report reveals how slavery and imperialism advanced epidemiological practices. The people of Boa Vista contributed to knowledge production. Their testimony not only assisted McWilliam in understanding the origin and progress of disease, but their participation in the study helped to solidify the interview as a fundamental method in epidemiological analysis.

✦ ✦ ✦

James McWilliam was sent to Cape Verde because he had experience studying disease in Africa. Born and raised in Scotland, McWilliam studied medicine at the Edinburgh College of Surgeons and entered the

Royal Navy, where he served for a number of years, including as surgeon aboard a navy ship patrolling the west coast of Africa. In 1840, after
returning to Edinburgh and completing his medical degree, he was appointed surgeon on the *Albert,* one of three steamships taking part in
the Niger Expedition, commissioned by the British government. The
expedition was a missionary and trade venture intended to thwart the
illegal slave trade by making trade deals with local leaders and establishing a model farm deep in the interior, along the Niger River. Several
weeks into the journey up the river, fever broke out on all the ships, and
the decision was made to return to England. With many officers and
crew members dead and sick, McWilliam himself had to navigate the
ship for a time. He, too, eventually fell ill but recovered after several
weeks of fever. Overall, McWilliam reported, of the 145 white men on
the expedition, 130 contracted the fever and 42 died. Of the 158 Black
people, 11 became ill but none died.[9]

Like many British military physicians of his generation, McWilliam
undertook scientific investigation while handling his medical duties.
After returning, he published his account of the journey and observations on the fever in a book titled *Medical History of the Expedition to
the Niger during the Years 1841–2, Comprising an Account of the Fever
Which Led to Its Abrupt Termination.* McWilliam's study of the Niger
Expedition provides important context for his later investigation of the
disease outbreak in Cape Verde. He developed epidemiological methods
that he would further sharpen in his study of Boa Vista.

The first section of McWilliam's detailed accounting of the voyage
contains descriptions of the ships' dimensions, regulations, outfitting,
and personnel; a complete itinerary of the ill-fated expedition; and observations on the geography, geology, climate, and anthropology of all
the locations visited, along with statistics on the men who became ill
and died, divided by rank and race. In Sierra Leone, where local people
were hired for the voyage inland, McWilliam marveled that "there
seemed assembled all languages, all shades of the dark race, marked after
the mode of their various countries." He and the captain chose inter-
preters from many nations: "Ibu, Kakanda, Haussa, Yoruba, Bornou,
Laruba, Eggarra, and Filatah." He also took interest in their spiritual
practices, relating, for example, the story of a couple who were struck

by lightning during a tornado, while they were beating drums and wor-
shipping the thunder. The African sailors and fishermen, he noted,
have many wives and "worship the devil" because, they said, God is
good but the devil is evil "and is therefore to be feared." Their doctors
cure people with "fetishes."[10]

He noted that most of the local people practiced circumcision, and he
remarked on a group of ten young boys, sons of one of the chiefs they encountered, who "had the hair cut close, or shaved, so as to leave it in tufts
and lines, describing diamonds and other angular figures over the head."
He included a sketch of one of the boys, as well as of a Nufi man from
Egga.[11] Although McWilliam's account offers ample evidence of white
supremacy and Orientalism—he refers, for example, to human sacrifice
and to "superstitious bigotry" in the local practice of medicine—his
report cannot be understood simply as yet another example of British
people portraying Africans as peculiar, uncivilized, and primitive.[12]
Many of McWilliam's observations are linked with his desire to uncover
evidence that might explain the outbreak of fever. Like many of his contemporaries, he focuses particularly on the climate and local environment.[13] He does not identify individual characteristics as causing
increased susceptibility to disease. Instead, like eighteenth-century
colonial botanists who traveled to the West Indies, his report suggests
that there is value in offering a panoramic view of West Africa in order
to provide a context for understanding disease.[14]

McWilliam discusses racial difference in disease outcomes, noting
that fever attacked most of the white people aboard the ship but produced mild disease in only 11 of the 158 Black people, "consisting of
natives of various parts of Africa, including Kroomen, Americans, West
Indians of African origin and East Indians." These eleven, who survived,
had all spent some years in England, "showing that the immunity from
endemic disease in warm countries, which is enjoyed by the dark races,
is to a certain extent destroyed by a temporary residence in another
climate."[15]

While McWilliam's claim about African people's innate immunity
might seem to indicate a focus on racial difference, he was more concerned with the influence of climate and environment, not physiology,
on the African people. In other instances, he discusses the Black

members of the expedition in the same way as the white members. In the chapter on "illustrative cases of the fever," which lists in detail the course of the disease in sixteen people, Case 2 is that of William Oakley, a thirty-eight-year-old man who served as a "gun-room cook" and was born in West Africa but had lived in England for several years. When Oakley first came under McWilliam's care, on the first day that fever struck on board, he had indigestion, was shivering and vomiting, and had a severe headache. Over the course of the next few days he had periods of fever and weakness. McWilliam took note of Oakley's bowel movements, pulse, the condition of his tongue, and how well he slept. Oakley returned to his duties over two weeks after becoming ill.[16]

In chapter 2 of his report, McWilliam summarizes his observations on the symptoms of the fever, effective treatments, and the results of eight autopsies he was able to perform. In addition to fever, common symptoms included headache, chills, fatigue, vomiting, sweating, and sometimes convulsions and delirium. For treatment, he recommended blisters, purgatives, and quinine, but cautioned against bleeding. During convalescence, removal to a cooler climate was the most effective; the fever would rarely abate "so long as the patient continues within the pale of malarious influence." McWilliam also concluded that the fever was not contagious, since several of the people who were in constant contact with the sick never became ill.[17]

While the sequence of symptoms helped to provide a clinical classification of the fever, McWilliam remained determined to uncover what caused it in the first place. "No subject in medical philosophy has been investigated with less satisfactory results," he wrote, "than the causes of those fevers which, in intertropical countries, are the especial bane of European life."[18] Like others at the time, he noted that the worst fevers occurred in lushly vegetated areas with marshes, frequent flooding, and high temperatures. It was generally believed that vapors resulting from decomposition of organic matter somehow led to disease.[19] McWilliam then entered into a long account of his experiments during the expedition on the subject of "sulphuretted hydrogen," or hydrogen sulfide.[20] Bottles of water collected off the coast of Africa by McWilliam himself, on a previous voyage, and by others, had been found to contain hydrogen sulfide, and the chemist John Daniell had proposed that there might be a connection between this gas and "the notorious un-healthiness of the coast of Africa."[21] If so, it might be possible for ships to test the waters and avoid locations where the gas was abundant, or it could be removed by fumigation with chlorine.[22] Medical officers on the Niger Expedition were instructed to collect and test water samples as they proceeded along the coast and upriver. McWilliam collected nu-merous samples but concluded that the gas that appeared within the collection bottles was an experimental artifact resulting from putrefac-tion within the bottle, and that the gas was not widely present in the environment: "I consider the absence of the gas in question from the sea and river waters and superincumbent atmosphere to be distinctly proved." That left the cause of the disease as the miasma, or "malaric exhalation," which was enhanced by stagnant air, high temperatures in the coal-fired ship, and the "enfeebled" and "dispirited" state of the men.[23] Despite the many failures of the Niger Expedition, it provided McWilliam with insights about the diagnosis, treatment, and causes of fever that would inform his study in Cape Verde.

❖   ❖   ❖

When McWilliam arrived in Cape Verde in March 1846, he followed similar investigative practices to those he employed during the Niger Expedition to study the spread of the fever. He had been ordered by William Burnett, medical director of the Royal Navy, to supply answers to a long list of questions. Was the disease endemic, epidemic, or spo-radic? Was it "indigenous" to the island, or could it have originated from a source unconnected with the *Eclair*? Did the women who washed the officers' clothes become ill, and "were they Africans or islanders? black or coloured?" Had fever attacked anyone in the houses where the offi-cers were stationed? McWilliam was, in addition, to find the names of all islanders who had visited or labored on the ship or the small island where the crew had been housed in the old fort, and of the guards sta-tioned at the fort, to determine if any of them had been attacked by fever, and if so, when; from these facts he was to determine "whether it ap-pears the disease was contracted from a morbific agent of a miasmatic nature, generated or existing within the vessel, or from exposure to a specific contagion emanating from the bodies of the sick." He was to ascertain whether the disease had been communicated to second and third parties—essentially, he was being asked to do a type of after-the-fact contact tracing.[24]

When McWilliam arrived, the outbreak was almost over. He was able to examine one or two patients, but to trace what had happened during the previous six to seven months, he relied on first-person testimony gathered from all over the island.[25] McWilliam's report begins with his interviews of the washerwomen who had received the officers' soiled linen from the British consul's storekeeper. The storekeeper named all seventeen women, and when McWilliam asked if he could bring them to the British commissioner's house to be interviewed, he replied that he could bring all of them except four, who had died of the fever.

By following the laundry trail, McWilliam hoped to answer a long-standing question about whether laundry, cargo, and other materials could be infected and spread disease. He transcribed his interviews, identifying each woman by her name, age, and racial group ("mulatto," "dark mulatto," or "negress"). The washerwomen were cosmopolitan and were careful eyewitnesses. One, Maria de Anna Limoa, reported that she had traveled to "America, Lisbon, Greece, St. Jago, and others of Cape de Verds."[26] Others, while perhaps spending most or all of their lives in Cape Verde, nevertheless would have met crews from different nations who visited the busy port on Boa Vista. All but one of the washerwomen did, in fact, contract the fever, and so did many members of their families. Before McWilliam arrived with his stack of paper, his quill pen, and his questions about the epidemic, these women had already taken note of the epidemic. They had spotted the unusual symptoms, timed when the illness appeared and how long it lasted, and kept track of who died and who recovered.

Limoa distinguished the fever that tore through the island from smallpox, which she had had when she was living on St. Jago, another of the Cape Verde islands. Emilia Joana Mariana had been ill, as had her husband and daughter. Her husband, a "country doctor," had visited many sick people and was the first in her family to catch the fever. Her daughter, who was staying with her grandmother in another town, had "black vomit" and died after four days. Anna Santa Anna, who identified herself as a slave, told McWilliam that when the fever first broke out, her master fled and took her with him to another village on the island. Once it dissipated they returned to Porto Sal Rey. Her master then took his daughter and another enslaved person to a different town—all of them became ill, but none who stayed in Porto Sal Rey, including Anna, did. This kind of information helped McWilliam track who got sick where, and when.[27] Maria Leonora, a "negress," told McWilliam that she contracted the fever around October 20, which was about six weeks after the *Eclair* had departed. Leonora said she was sick for a fortnight and that her mother and two sisters all became ill soon after she did.[28]

The washerwomen knew who was sick, who had been in contact with whom, and who had transported and buried the dead bodies, which provided important clues in tracing the spread of disease. Leonora told McWilliam that she had visited several of the sick people, including a man named Manoel Affonso. Another washerwoman, Antonia Chileco, described by McWilliam as a "mulatto," noted that her brother, Leandro Evera, who had also visited Affonso, had become ill and that he had been on a boat with Manoel Antonio Alves, who had buried two Portuguese soldiers, the first victims to die during the epidemic. Leandro and his family all became infected with the fever.[29] Antonia Romess, whom McWilliam identified as a "negress," said that her brother was the first in her family to have the fever. "He was taken ill after he had been carrying some dead bodies to be buried," she said, and died in November.[30]

McWilliam then interviewed the island's military commander and eleven of the soldiers, most described as "negro," who had served as guards at the fort where the crew members of the *Eclair* were housed. Five three-man groups of soldiers had guarded the fort in shifts during the time the crew was there. McWilliam, using information relayed to him by the surviving soldiers and townspeople, traced the initial outbreak of the fever in Porto Sal Rey, the main town on the island, to the movements, actions, and fates of these soldiers.

The first deaths among the guard were those of two Portuguese soldiers who had been stationed at the fort during the period when the crew members of the *Eclair* returned to the ship and left the island, on September 13. Many of the people whom McWilliam interviewed described hearing about these men's deaths and what happened afterward.

The two soldiers, who had been in and out of the sick room at the fort, became ill a day or two after the crew left and died after a few days. They were cared for by Miguel Barbosa, the third member of that shift, and Pedro Manoel, who was sent by the commandant to help take care of the sick soldiers. Barbosa and Manoel, both described as "negro," also cleaned out the sickroom the crew had stayed in, using gunpowder (thought to act as an air purifier) and whitewash. After the Portuguese soldiers died, another soldier, Manoel Antonio Alves, also described as "negro," was commanded to go to the fort and help retrieve the bodies.

Alves reported that when he arrived at the area near the fort, he docked his boat and then took off all of his clothes, presumably to prevent contamination. As he explained to McWilliam, he ran up to the fort "in a state of nudity," where he encountered Miguel Barbosa and Pedro Manoel. One of the Portuguese soldiers was already dead, and the other "was nearly dead." The three of them rolled up the dead body in a red quilt. "The body looked very bad," Alves told McWilliam, and "smelt most offensively." The men "put two pieces of board under the body," carried it to the boat, and took it a mile and a half away, where they "buried it in the sand on the beach." The next day the other Portuguese soldier died, and they also buried his body.[31] McWilliam interviewed both Miguel Barbosa and Pedro Manoel. Barbosa had witnessed many of the crew from the *Eclair* die while he was on guard duty and said that he and the Portuguese soldiers occasionally went into the sickroom. He described the soldiers' symptoms: "first, general fever; then delirium; and afterwards constant black vomiting." He told McWilliam that Alves had thrown the dead soldiers' clothing into the sea. Manoel confirmed much of what Barbosa said and also reported what happened after they left the island. The commandant, he reported, "fearing my comrade [Barbosa] and myself might have fever about us, did not allow us to return to the barracks at Porto Sal Rey, but put us both into a small house in Pao de Varella."[32] From that small house in a neighborhood of Porto Sal Rey, McWilliam traced the next stage of the outbreak.

McWilliam described the houses in Pao de Varella as "in general mere hovels, rudely built, much crowded together, and, with few exceptions, dirty." They were occupied by "the lowest classes, and the streets were filthy."[33] While housed there, Barbosa and Manoel both became ill.

They were attended by two women who lived in the neighboring houses: Anna Gallinha, a Portuguese woman who cooked for them, and Joanna Texeira, a "mulatto" woman. Numerous other people from the neighborhood also visited the sick soldiers.[34] A few days after Barbosa and Manoel returned to the barracks, Gallinha became ill. According to John Jamieson, the British storekeeper, she died within four days, showing symptoms of "high fever, wildness, and black vomit." Texeira served as her nurse, and a frequent visitor was another neighbor, Manoel Affonso, a laborer. Both got sick, and Affonso died. According to Texeira's testimony, and that of a few other townspeople, a number of others who lived in the same group of houses also died or caught the fever, as did the man who buried Gallinha and Affonso.[35]

In addition to the washerwomen, military guards, and townspeople from Porto Rey Sal, McWilliam also interviewed thirty-eight of the forty-one laborers who worked on the *Eclair* while it was docked (and who hailed from different parts of Boa Vista), eighteen of the twenty-three men who worked on the launches that fetched coal from an area near the fort and brought it to the ship, and twenty of the twenty-three who labored on the coal heap itself. These men provided details about when people came in contact with those infected with the fever and the amount of time it took for them or others to contract it. They also offered accounts of how the fever spread to different parts of Boa Vista. They identified when the fever reached their community. They described the symptoms. They counted the sick and named the dead. They provided dates and times, with more or less precision. In response to questions about when the fever arrived or how long it lasted, some responded by saying "a few days" or "a week," or "until the steamer came here," or "when it was common in my village."

McWilliam pieced together all of the testimonies to document the spread of the fever through Porto Sal Rey and across the island and how it was transmitted from person to person.[36] The washerwomen offered crucial details about transmission. Despite having direct contact with the soiled linen of the *Eclair*'s officers, they all emphasized to McWilliam that they did not become infected with the fever until weeks and sometimes months later, after the disease had become general on the island. Emilia Joana Mariana told McWilliam that she had had the fever, "but not until late in December last." Maria da Rocha said, "I was sick, but not until very lately." Sabina Diego reported being attacked "when the fever was general in town," after her husband became sick.[37] "I have examined the survivors," wrote McWilliam, "and have found that two were attacked late in October, five in November, two in December, three in January, and one not until some time in February. . . . So that in none of these cases can the occurrence of the fever be fairly attributed to infectious matter conveyed by the linen." He noted that "none of the washerwomen were among the first laid up in Porto Sal Rey," and the same was true of many of the laborers who worked on the ship. "Facts such as these," he concluded, "would indicate that the fever only propagated itself to those who approached others under its influence."[38] Although McWilliam was able to draw these conclusions from the testimony of the washerwomen, who themselves appeared to be aware that they had not acquired the infection from the soiled clothing, he did not explicitly state that it was the women themselves who helped him understand the spread of the epidemic.

The value of McWilliam's report for the historian lies less its analysis of the cause of the disease outbreak than in its emphasis on interviews, which became a core epidemiological method. McWilliam paid close attention to the fate and movements of enslaved people on the island and conducted interviews with many of them as well as learning about those who died. One who succumbed to the fever was Portajo, whose owner was João Baptista, acting vice consul in Boa Vista and owner of the salt works. Baptista had a large household consisting of forty-two people, including enslaved people. According to Miguel Barbosa and Pedro Manoel, Portajo was the only person who visited the Portuguese soldiers who died at the fort, other than a doctor. Portajo, who was cared for by two of Baptista's other bondspeople when he was sick, died sometime in mid-November. His illness appears to have prompted Baptista to move his entire household to his home in Boaventura, a district about four miles from the main port, where the fever had not yet appeared.[39]

Boaventura was located next to what McWilliam describes as a poorer district, Cabeçada, where the fever had started spreading as early as the end of September.[40] The residents of Boaventura tried to keep the disease away by isolating themselves: "We did not allow the Cabeçada people to come near our houses," testified Baptista.[41] But eventually the fever arrived. First, Baptista's brother-in-law died; he had been brought to Boaventura from Porto Sal Rey after falling ill. Then an ailing servant, Rosa Fortes, came to her father's house, and the entire household fell sick.[42] The third was an unnamed enslaved person. McWilliam asked Baptista whether he had heard about an enslaved person who had traveled during the night from Cabeçada to Boaventura, and Baptista said he had. McWilliam then interviewed Vicente Antonio Oliveira, whom he identified as a "mulatto, native, carpenter." Oliveira had heard about the death of Baptista's brother-in-law and Rosa Fortes. When asked if he knew of any other cases, he said, "Yes, a slave belonging to Cecilia da Britho went by stealth in the night-time from Cabeçada to Boaventura, to see a friend or relation of his; he was taken ill at Boaventura and died in four or five days of illness." The fever then spread, he said, "from Rosa's father's house, and the house where this slave died." The woman in whose house the enslaved person died, Ambrosia, also later died.[43]

McWilliam talked to a number of enslaved men who worked as laborers on the *Eclair* or on the launches used for restocking coal. He asked how many days they worked, if they had spent time aboard the ship or at the fort, whether they got sick and if so, when and where, and if they had been around other sick people. Their responses varied but, along with those of the other laborers, enabled McWilliam to determine how and when the fever spread.[44]

The interviews of the colonized and enslaved people on Boa Vista emphasized the importance of collecting first-person testimony about an epidemic in order to better understand it. McWilliam treated all of his informants equally and did not fall back on racial ideology to explain disease transmission. This is not to argue that racism did not inform or even shape how McWilliam and others interpreted this epidemic—he often referred to the people according to a rigid system of racial classification—but his report is a remarkable example of formally recording, documenting, and processing colonized and enslaved people's testimonies as central evidence in a disease investigation.

÷   ÷   ÷

Based on the interviews he conducted, McWilliam concluded that the *Eclair* did, in fact, transport fever to Cape Verde. He determined that

Cape Verde had been in a "healthy state" before the British vessel arrived. The fever that broke out on the island was the same one that was afflicting the ship when it arrived. But the nature of the disease had changed during its course on the *Eclair*. He argued that the fever when it first appeared on board the *Eclair* in late May and early June 1845 was "the usual endemic fever of the African coast, which . . . is not considered to be of an infectious or contagious nature." But during the *Eclair*'s trip from Sierra Leone to Cape Verde, the fever "changed materially for the worse," as evidenced by the fact that several of the cases on board exhibited black vomit, a symptom not found in the usual endemic fever. It "acquired still greater virulence" while the sick crew members were housed at the fort. "In short, from the endemic remittent of the African coast, the disease had, from a series of causes, been exalted to a concentrated remittent, or yellow fever."[45] His most important conclusion, from the point of view of the British government, was that this now contagious disease spread from the ship to the island of Boa Vista and was responsible for the fatal epidemic: "It is evident that the fever at Boa Vista possessed the properties which are usually attributed to a contagious disorder; and connecting this fact with the time and circumstances of the seizure of the soldiers at the Fort by fever, and of the appearance of the disease in Porto Sal Rey, will I think leave no doubt of its introduction into the island by the Eclair." From Porto Sal Rey, the fever was then "propagated throughout the island almost exclusively by direct intercourse with the sick."[46]

McWilliam suggested several causes for why the disease had become contagious. The situation on the ship in Sierra Leone was poor: the crew were engaged in "irksome" duties, and had by then been exposed to "morbific influences" for several months. Once the ship arrived in Boa Vista, the sick sailors were removed to the dilapidated fort, where McWilliam pointed to crowding and poor ventilation, the same culprits that Trotter, Milroy, and others had highlighted, as factors that increased the virulence of the disease.[47]

McWilliam believed that the high mortality on the island was due in part to poor nutrition as well as to what he described as "the total absence of medical assistance for some months," after the Portuguese doctor fled and the English doctor died.[48] He ignored the nursing by family members and the work of local healers, such as João Mariana, one of the laborers, who described himself as "a doctor of this country" and said he had attended "a great many" sick people.[49] McWilliam concluded with a list of measures to help prevent a recurrence of the highly infectious disease, such as dumping garbage into the sea, not allowing pigs to roam on the street, whitewashing the houses, and regularly airing bedding and clothing.[50]

McWilliam's understanding of yellow fever is by modern standards inaccurate. By the end of the nineteenth century, researchers had determined that yellow fever is transmitted by mosquitoes. But while McWilliam's focus on contagion might now be regarded as incorrect, his first-person interviews were remarkable. He recognized the patient's narrative as a central component to understanding the origin, character, and behavior of the epidemic. The systematic process of asking people within a community a standard set of questions became a fundamental practice in both public health and epidemiology.

McWilliam's *Report on the Fever at Boa Vista* was published in England and read throughout Europe and the United States, where it was reviewed in numerous journals.[51] Sir William Pym, superintendent of quarantine in Britain, supported McWilliam's assertion that the fever on the *Eclair* was contagious because it agreed with his views on yellow fever and validated his decision that the vessel be quarantined when it arrived in England (even though both Pym and McWilliam believed that yellow fever could not persist in cold climates).[52]

Burnett, who commissioned McWilliam's investigation, rejected the study because it blamed the *Eclair* for bringing the fever to Cape Verde from Sierra Leone. As a result, Burnett sent another doctor, Gilbert King, to conduct a second investigation.[53] King disputed McWilliam's conclusions and posited that heavy rain led to the outbreak of fever on Cape Verde and that it had nothing to do with the *Eclair*. He also questioned the reliability of McWilliam's informants, whom he described as "ignorant and illiterate persons in the very lowest grade of civilized society."[54]

To support his argument against contagion, Burnett brought up examples from Bermuda and Jamaica, where patients with yellow fever did not communicate the disease to anyone else.[55] Burnett, like McWilliam and other physicians, relied on case studies from around the world to understand epidemic disease. Yellow fever did not exist in England, but the reach of the empire enabled doctors like Burnett to receive reports from other parts of the world that informed their understanding of it.

McWilliam's report is also an example of how colonized and enslaved people's knowledge provided evidence to the physician, which then shaped how McWilliam's readers—doctors, government officials, and the public in London and throughout the Atlantic world—understood yellow fever.[56] Long before McWilliam arrived in Cape Verde, the inhabitants of Boa Vista had identified the symptoms and mapped where the epidemic spread. McWilliam became the authority on the subject of the epidemic when he returned to England and submitted his analysis as part of the quarantine debate, but his argument depended on the evidence that the colonized and enslaved people in Cape Verde had articulated. His questions made their knowledge visible.

Interpreting these people's insights as a form of knowledge production also allows us to recognize that describing details of illness and accounts of death to McWilliam was a form of emotional labor.[57] Luis Pathi, for example, a laborer described as a "dark mulatto," when asked "What family have you?" replied, "I have none left." His wife and three children had all died. His twelve-year-old daughter "had burning fever, and for several hours before she died, vomiting of black stuff and delirium."[58] Regardless of how devastating or disconcerting these scenes must have been, McWilliam's informants needed to revisit those details to answer his questions.

McWilliam interviewed over a hundred people, and the format of the interviews appears nearly the same for enslaved people, English merchants, Portuguese officials, and free Black people. He presents his question, and the interviewee offers their response. For this reason, the responses appear stoic and emotionless; the tone of their statements seems detached, whether they are describing their employment or the death of their children. The format of the interview offers no opportunity to capture a crack in the voice, a momentary silence, tears in the eyes, a disruption in the narrative. Instead, the interviews adhere to a rational scheme that shoehorns their statements into a framework that makes their knowledge legible.

The people who informed this study soon disappeared from the medical journals, newspapers, and policy debates as key informants. Yet their observations, insights, illnesses, suffering, and even their deaths contributed to how the medical profession grappled with the question of disease transmission, how policymakers debated quarantine restrictions, and how epidemiologists developed methods to investigate the spread of disease.

McWilliam's investigation is an example of how imperialism produced scientific knowledge, rather than how scientific knowledge was used to justify and propel imperial conquest.[59] That is not say that there were no power dynamics at work: the washerwomen, the enslaved people, the men who worked on the boats may have felt compelled to answer the questions, and they risked the possibility that their statements might mark them as the culprit who spread the disease. Power dynamics are also evident in the fact that the free Black and enslaved populations occupied the lowest forms of employment. They cleaned clothes and scoured out ships. They cared for the sick and buried the dead. They often became sick. Some even died, and many lost family members.

The *Eclair* controversy erupted almost a decade before John Snow began his famous study of cholera in London. When Snow entered Soho and talked to the impoverished residents, he was not the only physician taking this kind of approach.[60] McWilliam and other colonial physicians were undertaking similar studies across the globe. Many of them rubbed shoulders at the Epidemiological Society of London. Snow and McWilliam were both founding members of the society, which was established in 1850, and McWilliam served for many years as honorary secretary.

Another of the founding members, Gavin Milroy, became president of the society in 1864. Before that, he had served as a military physician in the British Caribbean. The significance of this region for epidemiological analysis begins with a story about a ship full of prisoners and ends with Milroy's analysis of the cause and spread of cholera in Jamaica in 1850.

CHAPTER 4

Recordkeeping

Epidemiological Practices in the British Empire

IN SEPTEMBER 1849, four years after the yellow fever outbreak in Cape Verde, a British convict ship, the *Neptune,* arrived in Simon's Bay at the Cape of Good Hope, South Africa. As soon as the ship sailed into the harbor, bells pealed through town to alert the residents. The people of Cape Town refused to allow the crew and prisoners to disembark. The butchers refused to provide them with meat. The bakers refused to offer them bread.[1]

Earlier that year, word had reached Britain's Cape Colony that plans were in the works to establish a penal colony in South Africa. Britain had been sending tens of thousands of prisoners to Australia for decades, but with resistance growing there, government officials began to look to other locations. The decision was made to transport hundreds of men who had been serving time in British prison hulks in Bermuda to the Cape.[2] When colonists in Cape Town heard that a ship full of convicts was on its way, thousands of them met and organized an Anti-Convict Association. Members signed a pledge that they would not allow the convicts to land or supply the ship with provisions.

When the *Neptune* arrived at Cape Town with 282 convicts, the townspeople honored their pledge and demanded that the ship depart. Their opposition made international news, where it was framed as an episode of resistance.[3] The *Bombay Times* lauded "the determination and the unanimity of the colonists," who refused to accept the "plague stricken" prisoners who would "infest" the colony with a "moral pestilence." The newspaper hoped that the townspeople's opposition would teach the convicts a lesson that "society abroad loathes the idea of their presence as bringing contamination along with it."[4]

The townspeople's campaign proved effective. The convicts never set foot in South Africa, and the British government finally gave in; the ship left for Van Dieman's Land (Tasmania) in February 1850, after five months at anchor. While this episode is well known, and is often framed as a case of colonial resistance, there is another, previously untold, part of the story.[5] As the British Empire grew, so did its bureaucracy. Ships' logs, surgeons' reports, official documents, and even letters and petitions from convicts now fill thick ledger books in the British National Archives. One of those letters is from a prisoner who was serving time in a prison ship in Bermuda, George Baxter Grundy, to Sir George Grey, British home secretary.[6] In the letter, dated May 1849, Grundy wrote that he intended to expose "the abominable wickedness carried on in the Hulks," and he asked Grey to launch an investigation. Grundy, who had been in the hulks for six and half years, detailed a number of abuses and incidents of mismanagement, and, in addition, reported that "unnatural crimes and beastly actions are committed on board the Hulks daily." He then described a world of same-sex intimacy, physical sex, and even marriage between men.[7] He reported witnessing men having sex on the ship; the convicts, he claimed, "boast" of this "abominable sin" and state that "if they are 'married' as they term it, it is out of the fashion." At least fifty men on the hulks kept "boys," whom they bought gifts for.[8]

Grundy's letter provides specific details about practices that had been widely rumored concerning convict ships. A report issued by a British select committee in 1837–1838, for example, had blamed the system of convict transportation for encouraging so-called corrupt vices, especially sodomy, which were often referred to using the same language of disease—"pestilence" and "plague"—that the *Bombay Times* used with reference to the convicts.[9] An appeal to the British government from the South African Auxiliary Bible Society decried the importation of convicts partly on the grounds that these "depraved" and "vicious" men indulged in "unnatural crimes."[10] Grundy's letter has been preserved as part of the same bureaucratic system through which observations about disease outbreaks around the world, full of details about individuals, flowed to the metropole. As a result of the British Empire's investigation of quarantine and contagion, fear of contamination was an increasingly common theme.

✛   ✛   ✛

Just as letters from prisoners flowed in to London, reporting on conditions in the empire's far-flung penal colonies, so did reports from physicians, documenting their observations, charting the spread of disease, establishing preventive measures, and devising theories that contributed to the development of epidemiology as a field. While Holroyd and other British military physicians had engaged in similar work in the early nineteenth century, Gavin Milroy, one of the founders of the Epidemiological Society of London, and others in the British Caribbean advanced these practices in the 1840s and 1850s. The expansion of bureaucracy and empire in the 1840s increased the number of records that were printed, circulated, and archived. Prior to the nineteenth century, doctors and learned citizens traveling abroad often sent letters to the Royal Society in London, where they might be read aloud in a meeting or published in the society's journal.[11] As the empire grew, a standardized system of recordkeeping developed that involved writing medical reports and sending them to medical and government authorities in the metropole.[12] This system formalized the process of turning observations into theories. Throughout the British Empire, physicians witnessed firsthand the outbreak of epidemics—from plague to yellow fever to cholera. Recordkeeping provided a framework for investigating, processing, analyzing, maintaining, preserving, and later archiving doctors' responses to epidemics. After returning to London, many of these doctors emerged as leading epidemiologists who published authoritative medical treatises and articles drawing on evidence they had collected from around the world, mostly from studying subjugated populations.

The cholera epidemics that ravaged London in the mid-nineteenth century are often regarded as the spark that ignited the emergence of epidemiology, but the spread of infectious diseases throughout the British Empire among formerly enslaved, colonized, and subjugated populations significantly shaped the field. Between 1817 and 1866, there were five separate cholera pandemics that began in India and then spread to Russia, Europe, and across the Atlantic to the Caribbean and North America. Unlike smallpox and plague, which had threatened communities for centuries, cholera was a relatively new disease for many Europeans and Americans. It horrified doctors and laypeople alike, with its violent symptoms of cramps followed by uncontrollable diarrhea and vomiting that emptied the body of all its fluid, killing the victim from dehydration within hours. Physicians did not understand what caused it or how it spread. They debated about whether it was contagious and what role quarantines played.[13]

Military and colonial physicians who encountered cholera and other infectious diseases abroad wrote reports about their observations. On ships, they collected information about patients and registered data into logs. In military camps, they produced weekly reports that tabulated the number of soldiers who became sick, who were admitted to the infirmary, and who died. In British colonies, they collected reports from local doctors and investigated outbreaks.[14]

Physicians serving throughout the British Empire did not simply theorize about epidemics but provided recommendations for how to treat patients and prevent further outbreaks. Because of their investment in the health of the public, they expanded their focus to include not just soldiers and sailors but also colonized people. As Gavin Milroy noted in a letter from Jamaica to the secretary of state for the colonies, if any means could be used to mitigate the cholera epidemic that was then devastating the island, "surely then it is the duty of all who have any influence whatever over the government, local or imperial, of a land, to see that these [means] are not wilfully overlooked, and that the people are not left to perish from mere neglect."[15]

Bureaucracy became a way to keep track of epidemics. Colonial physicians and administrators recorded where cholera spread, how many people it infected, and how many people it killed. They discussed symptoms and treatments. Bureaucracy offered a way for physicians to communicate with each other. These records offered order, providing a narrative for an epidemic disease that terrified and confounded people throughout the world.[16] A bureaucracy that had been established in the service of war, colonialism, and imperialism emerged as the foundation for the development of epidemiology.

Official reports from the far corners of the empire came in many forms, including the journals required of every naval surgeon, which included details on all diseases and deaths occurring on the ship; reports from physicians sent to the colonies for investigative purposes, such as Gavin Milroy to Jamaica and James McWilliam to Cape Verde; and reports submitted by colonial physicians. Some physicians, such as Thomas Trotter, wrote books when they returned home. These studies often discussed the same diseases but might use different terminology or propose different guidelines about treatment or theories about what caused the disease and how it spread.

٭   ٭   ٭

In 1847, while James McWilliam was investigating a mysterious yellow fever outbreak in Cape Verde, a British naval surgeon, James Henry, was observing cases of cholera among crew members on a steam mail packet, the HMS *Antelope,* in the Mediterranean Sea. In his journal, which included the required listing of illnesses and injuries taking place during the voyage, he reported two cases of cholera.[17] The first, which occurred in Constantinople in June, was mild, while the second, which occurred in September, two days after the ship left Malta, was more severe. Both men recovered. In the "surgeon's remarks" portion of his journal, Henry refers to these two cases and then discusses at length the cholera epidemic that was still going on in Constantinople when the *Antelope* arrived there in early summer 1848. He records his own thoughts on the questions everyone was interested in: what caused cholera and how contagious it was.

Based on information Henry had apparently acquired about the outbreak in the city, which began in October 1847, he concluded that the disease was not very contagious—the first few neighborhoods where it appeared, for example, were not near each other, and people in the same house did not always become sick. Echoing the prevailing view at the time, he attributed the outbreak primarily to poor sanitary conditions: "If we remember the great want of sanitary regulations at Constantinople, the ill ventilated houses, narrow streets, close bazaars, open sewerage, the cemeteries in the very centre of the living; every thing tending to spread a pestilential distemper, I think we are justified in taking it as a negative proof of the absence of an 'infectious propagation' of Cholera *such as it appeared at Constantinople in 1848.*" He also noted that the disease was primarily "limited to the lower orders," who experienced frequent "diarrhea and colica." He believed that those symptoms, rather than indicating a mild form of cholera, probably acted as "predisposing agents" and that they resulted from local eating habits.[18] "Any person who has witnessed the quantities of cucumbers, water melons, and fruits of various kinds consumed at Constantinople," he wrote, "will be perfectly satisfied that those diseases are likely to be of common occurrence." He likewise believed that although cholera was present in both Constantinople and Malta when the *Antelope* visited those two locations, the two occurrences on his ship "could be traced to irregularities of diet." Out of an abundance of caution, he isolated both patients, had their clothing and blankets disinfected, and ordered the bed of the sailor who suffered the more severe case to be destroyed.

Henry also noted that there was nothing peculiar about the weather that would "connect with the presence of the disease." As was typical of military and colonial physicians at the time, Henry kept copious, detailed notes about the weather in order to see if there was any correlation with disease. British bureaucracies systematized this form of data collection in the early 1800s, creating a government archive of scientific information.[19]

Like Arthur Holroyd, who collected information on the spread of plague, Henry was thinking about cholera on the level of the population and not just the individual. Being abroad in 1847–1848 gave him an opportunity that doctors in Britain lacked; the epidemic of 1831–1832 had ended, and the next one would not break out in London until September 1848. The two cases on his ship, as he noted, did not allow him to draw many conclusions, but in Constantinople he could observe a larger outbreak and also consult with physicians who had more experience with the disease. Physicians "of the Turkish Capital and at the Dardanelles," for example, told him that they had very good results from bleeding—though he was skeptical of their claim that they were able to reduce mortality to 10 to 15 percent.

Like other British physicians abroad, Henry took the opportunity to investigate disease outbreaks when he had the chance. He wrote up his ideas about the cholera epidemic in his "surgeon's remarks," which became part of the bureaucratic record. Naval surgeons were required to keep a journal as well as a daily sick book. They also had to file quarterly "nosological returns" that listed every disease and injury that had occurred on the ship, along with information on weather conditions, temperature, and any other factors that might have influenced health conditions. In Henry's "remarks" section, he wrote, before going on to discuss his ideas about the outbreak in Constantinople, "In the Nosological Returns for June I mentioned that it was my belief that the Cholera *as it then existed at Constantinople* did not present contagious properties; nor did it appear capable of transmitting itself by infection from the diseased to the healthy." Although he could not conclude much from the two cases on his ship, his observations in Constantinople enabled him to theorize more broadly and to enter his theories into the bureaucratic record.

Military and colonial bureaucracy that grew out of imperialism enabled physicians like Henry to see the spread of cholera across a broad geography. With epidemic diseases, the kind of information acquired from large-scale outbreaks, which played an important role in the development of epidemiology, was available to British physicians because of the global reach of the empire and the oppression of various populations.

❖   ❖   ❖

A series of cholera epidemics in the Caribbean between 1832 and 1853 resulted in numerous reports by British officials and physicians.[20] According to the British physician Gavin Milroy, who went to Jamaica during the epidemic of 1850–1851, the first reported outbreak of cholera in the Caribbean appeared in Cuba in 1833, with subsequent epidemics in the region occurring in 1849–1851.[21] Milroy believed that cholera had first arrived in the Greater Antilles from an epidemic that began in 1832 when European vessels docked in Montreal, Canada, and then spread south through New York and Philadelphia, eventually reaching New Orleans. From there it made its way to Cuba, where it decimated nearly 10 percent of the population, roughly 10,000 people, with high mortality especially among Black people. It then spread throughout the Caribbean and to Mexico. The next large outbreak began in 1848, appearing simultaneously in New York and New Orleans. It traveled north to Canada and, at the same time, appeared in the Caribbean and South America. Much of Europe was suffering from a major cholera epidemic at the same time.[22]

With the sudden outbreak of cholera in Jamaica in 1850, the British Colonial Office decided to send three medical inspectors to the West Indies. The lead physician in Jamaica was Milroy (whose work on plague is discussed in Chapter 2). Milroy, an ardent anticontagionist, had been a superintendent for the General Board of Health, when he had been involved in investigating and writing a report on the 1848–1849 cholera epidemic in England.[23]

Cholera was a particularly horrific disease because of its mysterious behavior, seeming unpredictability, and violent symptoms; it could seep into a community without any warning and leave people dead within hours. When cholera erupted in Jamaica, it led to massive mortality. According to some estimates, between 10 and 12 percent of the population died, which translates to roughly 30,000 to 40,000 people. The epidemic caused similar mortality rates in other parts of the Caribbean. Barbados lost roughly 13 percent of its population over three months, which amounted to about 20,000 lives. St. Kitts and Grenada witnessed similar fatality rates.[24]

The administrative structure of the British Empire and the practice of writing reports made it possible for Milroy to put together a narrative overview of cholera's movement throughout Jamaica. After he arrived on the island in January 1851, he began collecting information from physicians and military officials and also traveled across the island himself. As he explained in a letter to the undercolonial secretary, "I left Kingston on the morning of the 1st instant, and reached this town

[Lucea] this evening of the 8th, having passed successfully through the parishes of Dorothy, Clarendon, Manchester, Elizabeth, and Westmoreland. On my journey to this town, I have visited several places where the Cholera has prevailed with greater or less force, keeping an accurate account of all the circumstances which have appeared to influence its development, and the evidence of its attacks."[25]

The ability to travel enabled medical inspectors like Milroy to obtain a broad geographic overview of disease outbreaks. In London and New York City, most doctors' studies were geographically bound to the municipality in which they lived.[26] As Milroy explained in the introduction to his report on cholera in Jamaica, "Having had the opportunity of visiting every parish in the island, and of coming in contact and conversing with persons in all conditions of life, the field of my observations and inquiries was at all events abundantly ample. Moreover, the correspondence which I have kept up with some of the best informed residents since my return to this country [England] has furnished me with fresh evidence upon many points, on which that collected by myself was less complete than I desired."[27]

In his report, Milroy offered case histories of patients to illustrate how quickly the disease could progress to death. He described the case of a Black fisherman named Phipps, who lived near the "filthy negro yards" and left his "hovel" early in the morning to go to Port Royal. During the day, he was "attacked with diarrhoea and pain in the bowels." Later, after returning home, he felt "faint" and "very weak." During the night he vomited and continued to have diarrhea. The next morning and afternoon, the symptoms persisted. He became sicker during the night and died the following morning. His was thought to be the first case in Kingston.[28]

At the time, in England, Europe, the United States, and other parts of the world, a fierce debate was raging over the question of whether cholera was contagious. Milroy noted that the woman who lived with Phipps did not contract cholera, nor did any of the other people living nearby.[29] Phipps and his companion provided Milroy with evidence for larger medical and political debates.

Immediately following that account, Milroy detailed how the disease spread in the Kingston penitentiary. Shortly after cholera first broke out in Jamaica in mid-October, the governor authorized the use of one hundred convicts to clean the streets. Two days after the men had first begun working, one of them became sick and died that evening. Three days later, another prisoner died. During the following week, twenty-three more died. The epidemic continued to spread throughout the prison, which was populated almost entirely by Black prisoners (of over five hundred prisoners, seven were white and twenty-six "brown or coloured"). Within a matter of weeks, 128 of 508 prisoners, nearly one quarter, had died. Milroy presented a table showing the dates and numbers of prisoners who were attacked and died, and the number of days from initial attack to death. He noted that most of the men died at night, when the ventilation was "very insufficient."[30] Like many other physicians throughout the world, Milroy used the prison as a site for understanding infectious disease.

Milroy also collected information from military officers on other Caribbean islands, including Barbados, Dominica, and St. Vincent, where cases of cholera appeared among the troops before the outbreak in Jamaica. He summarized the information these officers provided on distinguishing symptoms of dysentery and other bowel disorders from cholera, recovery rates, and other characteristics, such as the fact that in Barbados, "all the black troops remained nearly exempt." Through such reports, physicians were able to share information and develop a better understanding of cholera.[31]

Military and colonial physicians' experiences in other parts of the world provided them with the opportunity to recognize cholera. When cholera broke out in places like London or New York City in the 1850s, some physicians had seen it during the previous epidemic of the 1830s, but many had not. As James Watson, the surgeon at the naval hospital in Jamaica, wrote in an 1851 article on the epidemic, "The symptoms in the persons attacked here were similar to those which I witnessed in Lisbon in 1833, and which are reported to exist in cholera patients in all parts of the world." Noting that the "medical history" of the disease, including symptoms and treatment, was already well known, Watson stated that his goal was "to throw as much light on the statistics of cholera in Jamaica, on this its first visit to the colony, as my opportunities of observation and my abilities will permit me to do."[32]

Military physicians were interested in understanding how cholera could spread from one location to the next and added their observations to the ongoing debate about the cause of the disease and mode of spread. Watson noted the details of the spread of the disease in Jamaica, in part by discussing its relation to the various ships that arrived from Panama and Nicaragua before and during the epidemic. Many Jamaicans, noted Watson, believed that the disease was contagious and that the outbreak could be traced to two brothers who had arrived on a steamer from Panama a week before the first reported case of cholera in Port Royal. The two men reported that their father had died of cholera in Panama, but they had no signs of disease other than "common intermittent fever." They recovered, under Watson's care, after taking quinine. Within a month of this case, cholera spread throughout the Black community of Port Royal, where nine physicians were deployed to treat patients who lived in "fetid hovels." From there it spread to Spanish Town and Kingston. Watson observed that among the "respectable civilians," mortality was higher in these towns than it was in Port Royal, perhaps because residents of Port Royal, he suggested, were younger and more vigorous.

Watson disputed the hypothesis of contagion by referring to the fates of several ships that arrived in Jamaica during the epidemic. One ship had arrived from Nicaragua before the outbreak, and thirty crew members were in the hospital suffering from intermittent fever; only one died of cholera. A second ship arrived from Nicaragua five days into the outbreak; over one hundred crew members were sent to the hospital with fever, and twenty-two died of cholera. At the same time, seven members of the hospital staff died. Finally, a third ship from Nicaragua arrived when the epidemic was almost over. Many of the crew were suffering from intermittent fever and were taken to the hospital, where they were treated in the same wards as the cholera patients and were attended to by the same medical staff, but did not contract cholera.[33]

Watson used this evidence to argue that cholera does not spread by human contact. Of those who claimed that cholera had spread through Jamaica by contagion, he asked, "Will they be pleased to explain how it happened that, if this contagion be so virulent, fifty highly predisposed men were thrust into hospital wards which were still reeking with the emanations from the bodies of cholera patients, and that not one of them took cholera?"

Watson not only rejected contagion theory but suggested that belief in it among the Black people in Jamaica contributed to their suffering and hindered them from caring from each other. He claimed that in Kingston and Spanish Town, the result of "teaching the people to believe cholera is contagious" was great panic: "Husbands refused to put their hands on the dead bodies of their wives to lift them into coffins, and even mothers deserted their children when the latter took the disease." In Port Royal, by contrast, where "we preach" that cholera is not contagious, "the poor people showed no unwillingness to help one another in their deep distress."[34]

+ + +

In the lengthy report that Gavin Milroy submitted about his observations of the cholera epidemic in Jamaica, he concurred with James Watson's conclusion that cholera was not contagious, and he carefully tracked the outbreak and spread of the disease. The first case in Jamaica, he reported, citing Watson, was that of an older woman named Nanny Johnston who lived in Port Royal. There was a rumor that she had washed the clothes of the two brothers who were accused of bringing the disease from Panama, but it was false, claimed Milroy. The woman who did wash the brothers' clothes got sick, but not until two weeks later, when the epidemic was already widespread.[35] From Port Royal the disease spread quickly across the island, creating devastation as it went.

Milroy described terrible scenes, as fear became part of cholera's pathology. In Kingston, he wrote, carts overflowed with coffins, and graves could not be dug quickly enough. He told of one case in which a person's body was found in a gully, being eaten by vultures. The victim's family, he suggested, had thrown the body there "either from the dread of contagion or to avoid the expense of the burial." It took a "large bribe," he added, to convince a group of "coolies" to dig a hole for the body. Heading east from Kingston, the disease was particularly virulent in Yallahs Bay, where "the people died like rotten sheep" and corpses were left for dogs and vultures to prey on.[36]

While the death toll was alarming, it provided British doctors with an opportunity to investigate the epidemic. Colonialism created an infrastructure that enabled physicians to begin to observe cholera outbreaks in many regions at the same time, which allowed them to undertake epidemiological studies of populations and environments.

The function, format, and genre of colonial recordkeeping was in dialogue with sanitary reform efforts in the metropole. Edwin Chadwick, the leading sanitary reformer in England, collected medical officers' reports about sanitation and disease around Great Britain and used them to produce his influential *Report on the Sanitary Conditions of the Labouring Population of Great Britain* (1842).[37] Milroy wrote to Chadwick in 1851 about the devastating mortality rate in Jamaica. "I am not in the least surprised at the dreadful ravages of the pestilence in many places," he wrote, "for example nearly two thirds of the inhabitants of a town, and four-fifths of the Negroes on a sugar estate, or in their own settlements. A prodigious deal requires to be done in the way of improvement both in the towns and in the country." Milroy noted that he had requested copies of Chadwick's 1842 report and the 1847 report of the Metropolitan Sanitary Commission to be sent to Jamaica in order to assist physicians in their fight against disease: "It appears to me nothing but fair that such valuable public documents bearing so intimately on the social welfare of the people in all countries and climates, should be widely circulated in our own colonies and be in the hands of almost every Medical Practitioner."[38] Milroy's request for the British reports on sanitary reform illustrates how information flowed from the metropole to the colonies through bureaucratic channels.[39]

The same colonial structures that allowed ideas about sanitation to crisscross the Atlantic Ocean had also created the structural inequities that left Black people suffering on plantations throughout Jamaica without medical care.[40] Colonialism led to the division of Black people into those who lived on "their own settlements" and those who worked on "sugar estates." Colonialism hindered their movement and tethered them to particular areas where cholera thrived, which caused Black inhabitants to die in larger numbers than white inhabitants of Jamaica. Milroy noted that Black people lived in places where there were no physicians or, at most, one physician to attend to thousands of people. As he wrote to Chadwick, "Large districts and parishes containing 12 or 15 thousand inhabitants, and 15 or 20 miles in extent have been left to the care of one practitioner. The result has been that a vast majority of the cases were not seen at all, and death has been sweeping off thousands almost quite unopposed."[41] In part, this resulted from the inconsistencies of the colonial bureaucracy to report the health conditions of colonized populations only when an epidemic struck. As Milroy explains, "While, in the case of the military and naval services, there have always been at home special Boards of superintendence and direction, to whom periodic reports of the health of our soldiers and sailors have been regularly sent from every station in the world, and who are thus enabled to devise and carry out suggested measures of improvement, there has been no machinery whatever for even so much as ascertaining the condition of the mass of the people in our colonial possessions."[42]

In his letter to Charles Grey, governor of Jamaica, which accompanied his report, Milroy detailed the housing conditions that Black Jamaicans had to endure. He noted that the lack of ventilation exacerbated the spread of disease and, like many other eighteenth- and nineteenth-century reformers, he used the British prison system for comparison. "In recently constructed prisons 1,000 cubic feet are allowed to each inmate," he wrote, "and it is an instructive fact that most of these buildings at home escaped the visitation of the cholera, although in many instances it was raging all around them, while, on the other hand, the most terrible ravages took place in other public institutions where the inmates were unduly crowded in badly ventilated dormitories." He described the dwellings of the Black residents in Jamaica as very poorly ventilated: "Generally situated in the very worst localities, squatted down upon the bare earth, without, in most instances, even a few boards for a flooring; choked with rank vegetation close up to the very door, and almost always surrounded with filth and refuse, little is the wonder that the pestilence committed such devastation among the occupants." Six to ten people might be crowded together at night in a small room with the windows and doors closed. "The state of the air in such a place at night,—the very time, be it remembered, when the choleraic poison is most active, and when the vast majority of seizures occurs,—is so nauseously close and oppressive that a visitor can with difficulty remain in it for even a few minutes."[43]

In his letter to the governor, written after he had completed his tour of the island, Milroy presented his conclusions about the factors that had led to the outbreak.[44] While declining to speculate about how cholera had first been introduced, he listed a number of local circumstances that made the disease "so virulent and fatal in Jamaica." The most important was "an impure or contaminated state of the atmosphere," arising from decomposing material and human respiration: the disease had hit most hard, he claimed, in locations that were "filthy and neglected," as well as in those that were overcrowded and poorly ventilated. Dampness and climate were also contributing factors, he believed, pointing to the unusually wet weather that had preceded the epidemic.[45] Finally, the lack of physicians meant that people were not treated at an early stage of the disease, when they might have been healed.

Milroy laid out a number of recommendations, which included increasing the number of physicians, appointing a medical officer for each parish, establishing local boards of health, and keeping better mortality records. Sanitation was of the utmost importance: "Cholera is the most unerring inquisitor, as well as the most fearful avenger, of sanitary neglect."[46] In his report, Milroy listed a variety of sanitary improvements that should be undertaken: garbage and other decaying matter should be promptly destroyed; sewage either removed or buried and covered; dwellings enlarged, ventilated, and built on high ground; and bodies buried away from town.[47]

Milroy drew particular attention to the water supply. In his letter to Chadwick, he wrote, "One town here, Falmouth, is exceedingly well supplied with water from a river two or three miles distant . . . this has been the case for the last 40 or 50 years. Two years ago the people of Kingston began to have something of the same sort, but the supply is still imperfect there."[48] "The water supply in Kingston," he wrote in his report, "was formerly derived altogether from wells in the streets and in private yards. The water, thus obtained in the lower part of the town, is apt to be brackish and otherwise impure. It is no uncommon thing to find wells sunk in the immediate vicinity of huge unbricked privies, whose fluid contents readily permeate the loose soil." Although a company had been formed to bring water from the mountains, the project had been poorly engineered, in Milroy's view, and because no drains had been provided, dirty water accumulated in the streets when they were cleaned.[49]

✦ ✦ ✦

The outbreak of the cholera epidemic in Jamaica reflected the messiness of a postemancipation society. Before emancipation, many planters had taken some responsibility for enslaved people's health, but after slavery ended in 1833, a system of apprenticeship emerged that became what the historian Thomas Holt refers to as "a halfway covenant." The relationship between planter and worker continued in the pattern of slavery in that the master was in charge of the enslaved person's labor for "forty and one half hours of the work week," but during the remaining time, the relationship resembled that of employer and employee: both parties worked out the labor assignments and negotiations. After the apprenticeship system ended in 1838, planters charged freedpeople rent to remain on their plantations. While some freedpeople paid to remain on the land, since it was close to their employment, many eventually left and established settlements in the mountains, where they no longer depended on the planters for their survival.[50] At the time of the epidemic, planters were trying to lure freedpeople to return to plantations to work with the promise of providing them with medical support, but many freedpeople had already created their own settlements. In these locations they had autonomy over their labor, but these makeshift communities were vulnerable to epidemics because of inadequate infrastructure. The cholera outbreak, in turn, propelled British authorities to institute public health measures, which varied across the British Caribbean from the establishment of a Public Health Act in 1856 in Barbados, to the establishment of garbage disposal in Grenada, to concern about providing clean water in Jamaica.[51]

Although Milroy recommended many improvements in his report, he was frustrated that the assembly in Jamaica was slow to enact legislation, as he wrote to a colonial administrator: "There has been, I regret to add, a great apathy of concern among almost all classes but especially among the members of the Legislature about the objects of my mission notwithstanding the late dreadful disaster to the island. With the exception of one or two members, the House of Assembly seemed to think that next session would be quite early enough to consider the propriety of enacting sanitary and medical relief measures."[52] Many of the measures he suggested were not put into place until years later.[53]

Even though Milroy's suggestions were not taken up right away, many of them were measures that would have helped to reduce the spread of cholera. John Snow's 1854 study of cholera in London is popularly regarded as the foundation of modern epidemiology, because Snow determined that cholera was spread through water contaminated by sewage. But Snow was only one of numerous physicians studying cholera, and his ideas were initially dismissed. After Milroy presented his findings from Jamaica at the Epidemiological Society of London meeting in December 1851, John Snow brought up his own theory that sewage in the Thames caused cholera to spread; he was laughed at for "favouring the absurdity of infinitesimal doses."[54] From James Henry on the *Antelope* to Watson and Milroy in Jamaica, physicians around the world were observing and making theories about the disease. Although in retrospect, Milroy's belief that cholera was communicated through contaminated air did not hold up, the importance he and others placed on clean water as part of sanitation improvements was crucial for disease reduction efforts. Much of this medical knowledge originated, circulated, and became codified as a result of colonial and military bureaucracy.

Milroy asserted that his findings with respect to "the influence of putrescent effluvia" on the occurrence of cholera corroborated the conclusions of the British General Board of Health. He was able to draw this conclusion by investigating the presence of cholera in various towns throughout Jamaica, observing the sanitary conditions, and evaluating Black people's health. Once cholera had dissipated in England in 1849, doctors needed other places and people to study. Milroy's study of cholera in Jamaica combined with reports of other physicians stationed throughout the British Caribbean offered additional case studies for investigating the disease. He also used his findings to continue a broader global discussion about quarantine.

After eight months in Jamaica, Milroy returned to England. In 1853, he was elected a fellow of the Royal College of Physicians.[55] In 1855, he was appointed as a member of the Sanitary Commission to be sent to

Crimea by the War Office. In 1864–1865, he served as president of the Epidemiological Society of London. Many of the founders and active members of the society, like Milroy, had experience abroad. This society, as Milroy stated in his presidential address in 1864, aimed to investigate the cause and prevention of disease in all countries and climates. In focusing on prevention, the society differed from the Royal Medical and Chirurgical Society, which focused chiefly on curing disease and relieving suffering. Milroy noted that the Epidemiological Society's appointment of foreign and colonial secretaries "recognised most emphatically the importance of obtaining wide and extended inquiries in different regions of the earth." With reference to the continuing questions about cholera, he concluded that much more information was needed, but that "no country in the world possesses such extensive opportunities for the observation and notation of all natural phenomena as Great Britain, in consequence not only of her numerous colonial possessions, but of her wide-spread consular machinery, whereby almost every foreign land is brought within the regular cognisance of the government."[56]

÷ ÷ ÷

Many of the epidemiological studies that took place in the mid-nineteenth century can be traced to military and colonial bureaucratic records that detail the work of physicians who were investigating the cause, spread, and prevention of disease in various environments across the world. While the British Crown sent physicians to the Caribbean to protect its economic investment, this endeavor had the unintended consequence of advancing the development of epidemiology. Reporting about how to avert the outbreak of disease and how to arrest its spread—a key hallmark in epidemiological practice—was an important element of British imperialism. As Milroy explained in the summary that accompanied his report, "Preventive measures are much to be preferred to curative or remedial ones."[57]

Military and colonial bureaucracy, in turn, functioned as a subregime of knowledge production.[58] While the publication of books and journal articles combined with presentations at professional societies, academic conferences, and universities codified ideas, bureaucracy served as a crucial albeit overlooked arena of knowledge production.[59] The writings of physicians on ships and in colonies provided a platform for British physicians to exchange ideas, to create professional networks, and to learn about how others theorized and practiced medicine.

Military and colonial bureaucracy, and the required recordkeeping, contributed to the expansion of epidemiology, but it also captured a range of details about everyday life in the nineteenth century, including same-sex desire and intimacy aboard prison hulks in Bermuda, and the living conditions of formerly enslaved Jamaicans. Because colonialism depended upon communication from the metropole to the colonies, reports continually flowed from the Caribbean to England to other parts of the world that documented the ordinary iterations of everyday life. Epidemiological practices depended upon the proliferation of reports, the circulation of knowledge, and the subsequent creation of archives that preserved this information.

Bureaucratic medical records provide us with a more personal, intimate understanding of the history of epidemiology. These reports, letters, and journals highlight physicians' questions and doubts, their insecurities and uncertainties, their fear about the spread of epidemics. They also reveal profound, poignant, and personal details about how ordinary people—including the enslaved and the colonized, the subjugated and dispossessed—reacted to epidemics. Details about their living conditions, family life, and work conditions appear in bits and pieces in these records. Though it's an incomplete, fragmented portrait of their lives, these sources reveal how people in colonized lands were instrumental populations whom physicians studied. They appeared first as evidence of a disease outbreak of epidemic; then they appeared as witnesses to how it spread and as vulnerable communities in need of preventive measures. As ideas became codified into recommendations about ventilation, clean water, and other sanitary measures, these people disappeared from the page.

The theories, principles, and practices that underpin epidemiology were based on people whose suffering, sickness, and, at times, death contributed to the advancement of the field. Colonialism created the conditions that made large groups of people around the world available for study and contributed to the development of the methods of epidemiology—identifying the cause of disease, tracking its spread, devising preventive measures, and developing a network of physicians, reformers, and government officials.

The same year that Milroy drafted his report on cholera in Jamaica, the Crimean War began. The outbreak of the war accelerated the development of epidemiology, especially through the work of Florence Nightingale.

CHAPTER 5

Florence Nightingale

The Unrecognized Epidemiologist of the Crimean War and India

I N THIS CHAPTER WE SHIFT TO WAR, which, in the mid- to late nineteenth century, enters as a major force that created large populations available for the study of infectious disease. Military physicians turned their attention to the medical crises caused by war—sick and dying soldiers, unsanitary camps, disease outbreaks. As in the case of slavery and colonialism, these biological catastrophes led to the proliferation of reports on the cause, spread, and prevention of disease. The story of wartime medical innovation is not unfamiliar, but historians have typically separated war from colonialism and slavery despite the fact that they were all occurring at the same time.[1] The advances that military doctors made during slavery and the expansion of the British Empire between 1755 and 1853 affected how medical authorities on the battlefield documented, interpreted, and understood the spread of disease. Similarly, wartime medicine produced important studies that advanced the field of epidemiology and shaped how British and American physicians understood disease causation and transmission.

Wartime medicine also alerted the general public to the problems of crowded spaces and unsanitary environments. These concerns, which had been circulating primarily among medical and government authorities, began to appear in newspapers, which reported on the gruesome suffering and appalling conditions within battlefield hospitals. The Crimean War (1853–1856), in particular, drew attention to the deplorable conditions of military hospitals and, as a result, to those of civilian hospitals in England and the United States.[2]

During the nineteenth century, from Europe and the United States to the far stretches of the British, French, and Spanish Empires, many people viewed hospitals primarily as institutions for the poor and dispossessed.[3] Most people preferred to be treated by physicians or healers in their homes; in fact, most people were born and died at home.[4] Those who did not have family members or a communal network to take care of them were often forced to enter hospitals, which functioned more like contemporary soup kitchens or homeless shelters.[5] Even though local governments controlled some of these institutions, they were woefully underfunded and understaffed, and struggled to provide medical care.[6] Despite some reform efforts, such as those of John Howard in the late eighteenth century, little was done to improve hospital conditions, and few recognized how these institutions, designed under a banner of benevolence, exacerbated illness and mortality. It was medical reformers, most notably Florence Nightingale, who produced persuasive evidence documenting the ways in which military hospitals, like prisons, slave ships, and colonial plantations, fostered the spread of disease.

While it was widely known that conditions in hospitals produced more harm than good, systematic evidence was needed. Without a way to track how many people became sick while hospitalized or died from a disease unrelated to their admission, specific stories of illness or abuse were easily dismissed.[7] The outbreak of the Crimean War created an unprecedented opportunity for sanitary reformers to gather statistical evidence to document the dangers of hospitals.

÷   ÷   ÷

British journalists were the first to expose the problems of the military hospitals. When the *Times* of London deployed William Howard

Russell to cover the Crimean War in 1854, he emerged as one of the world's first war correspondents. He advocated for more doctors to be sent to the region and blamed the military for mismanagement and for neglecting injured and sick soldiers. "With our men well clothed, well fed, well housed . . . there is little to fear," he wrote, at the beginning of the campaign. But, he warned, not providing these basic necessities could be disastrous. He reminded readers that during the Russo-Turkish campaign in 1828–1829, "80,000 men perished by 'plague, pestilence, and famine.'" To prevent "a repetition of such horrors," better medical care was essential: "Let us have an overwhelming army of medical men to combat disease." Russell described the hospital at Gallipoli as unprepared to receive sick patients: "They had neither bedding, covering, medical stores, nor medical comforts." Sick soldiers were given only a single blanket.[8]

British readers followed his dispatches and gained firsthand knowledge about the conditions of war, which had been obscured by patriotic narratives of valor and honor. His accounts reported on conditions that had previously been limited to military, medical, and government officials. As the *Times* published more of Russell's articles about the hospitals' ghastly conditions, the public grew angry and demanded reform. Russell, in turn, addressed his audience directly and not only called for an army of doctors but also pleaded with women readers to help the war effort: "Are there no devoted women among us able and willing to go forth to minister to the sick and suffering soldiers of the East in the hospitals in Scutari? Are none of the daughters of England, at this extreme hour of need, ready for such a work of mercy?"[9]

In early October 1854, Florence Nightingale, who served as superintendent of the Hospital for Invalid Women at Harley Street in London, read articles by Russell and other *Times* correspondents on conditions in the Scutari hospital. There, the men "expire in agony," without "the hand of a medical man coming near the wounds," and "there is not even linen to make bandages for the wounded."[10] After reading the descriptions of the terrible conditions in the hospital, she immediately wrote to her friend Elizabeth Herbert, who was married to the secretary of war, Sidney Herbert, about going to Scutari to serve as a nurse. She explained to Mrs. Herbert that "a small expedition of nurses" had been organized to go to Turkey to help the "wounded wretches." Nightingale said that she had raised private funding "to feed and lodge ourselves" and that she and the other nurses would not cost the country any money.[11]

At the same time that Nightingale was writing to Herbert, he was apparently writing to her. When the war first began he had suggested developing a corps of women to serve as nurses, but military and government authorities had opposed his proposal. After Russell's articles in the *Times* generated a public outcry about the conditions in the military hospitals, the government agreed to allow Herbert to organize a group of nurses to be sent to Turkey. Herbert, a reformer who had already established efforts to improve the health conditions of the army, believed that Nightingale, whom he had met several years earlier, had the credentials to lead the expedition.[12]

On November 4, 1854, a mere three weeks after Nightingale wrote to Herbert, she arrived at the Barrack Hospital in Scutari, on the Asian side of the Bosporus near Constantinople. Herbert provided her with strict instructions to follow the direction of the chief army medical officer at Scutari.[13] Since it was unprecedented to send women to the warfront, there was no protocol to guide Nightingale and the other women's work there. Nightingale was put in charge of hiring the nurses, determining their duties, and establishing their schedule. She also worked as the "barrack's mistress," which involved organizing the cooking, laundry, and distribution of supplies.[14]

Nightingale arrived in Scutari about a week after the Battle of Balaclava and the Charge of the Light Brigade, a famous battle in which the Light Brigade was sent to the wrong location and the Russians defeated them, resulting in high British mortality. Russell reported that the cavalry and artillery lost 175 men, including 13 officers, and that an additional 251 men were wounded, including 27 officers. Russell underscored the soldiers' bravery and heroism, and noted that the "emergency" that faced the soldiers was not the enemy on the battlefield, but disease. In a lyrical turn, he described how the soldiers had been "transported at once to bear the heat of a semitropical sun and the miasmata of a pestilential climate," where they "have seen, week after week and month after month, their comrades dropping around them under the stroke of an enemy that walks in darkness, against whose shafts the valour that can stem the tide of battle or scale the deadly breach can avail nothing."[15]

On November 25, 1854, in a letter to Herbert, Nightingale emphasized the problems of bathing and laundry in the hospital, which housed over 2,300 men. She complained that the hospital purveyor did not believe that washing was important. Thirty men per night were bathed, which meant that each man was washed only once every eighty days. The consequences were "Fever, Cholera, Gangrene, Lice, Bugs, Fleas,—& may be Erisypelas—from the using of one sponge among many wounds." When she arrived, she found no sinks, soap, or towels in the wards. She and her staff set out to collect and wash both the "body-linen" and the "bed-linen" and to create "a little Washing Establishment of our own."[16]

In early February 1855, after months of attempting to improve the health conditions in the hospital, Nightingale penned a letter to her mother in which she stated that despite the challenging obstacles, she remained committed to reforming the military hospital and the Army Medical Board. She reassured her mother that her education had not been wasted and that the principles that she had learned would continue to guide her work as a reformer. Her parents had initially opposed her plans to become a nurse, so this seems to have been her way of justifying the importance of her career. She even compared her reform efforts to the work of Christ.[17]

Three weeks later, Nightingale wrote to Sidney Herbert that conditions in the hospital were improving. She noted that they had buried only 10 men in the previous twenty-four hours, compared with 506 during the first eight days of February.[18] Although historians have debated the question of whether Nightingale's work lowered the mortality rate in the hospital during the winter of 1854–1855, there is no question that she publicized the dangerous and unhealthy conditions within military hospitals, which had been largely ignored before the war.[19]

✣   ✣   ✣

Like Gavin Milroy and other medical professionals who worked abroad, when Nightingale returned to England, she published reports that presented her findings. In her treatise aptly titled *Notes on Hospitals,* she described the filthiness that plagued both military and civilian hospitals in England and Europe. In response to a question from a royal commission on the sanitary condition of the army, which had been established in 1857, she wrote that when she arrived at the hospital in Scutari, in addition to the dirty walls and ceilings, "the rats, vermin of all kinds, accumulations of dirt and foul air, which harboured under the wooden divans on which the men were laid, rendered the atmosphere still more dangerous to them."[20] Stories of rats in hospitals were not uncommon; they often appeared as a trope to indict dirty hospitals. Across the Atlantic, in the United States, the popular magazine *Harpers Weekly* featured an illustration showing rats crawling on a woman patient asleep in her bed in an asylum. In the mid-nineteenth century, before rats were identified as carriers of plague, hospital staff may have considered them a nuisance but did little to expel them. Nightingale, like other reformers, described their presence to illustrate how dirty the hospital was.[21]

While the popular narrative of Nightingale focuses on her important duties as a nurse, leading the first-ever corps of women to the battlefield and treating wounded soldiers, her contributions are even more significant. The iconic image of Nightingale during the Crimean War features her as "the lady with the lamp," walking among wounded soldiers at night in the hospital ward while holding a lantern. This nickname and image conform with the military's expectations about her work in Scutari. That she was a "lady" who assisted physicians in the medical relief of the patients fits into a larger story of how her indefatigable efforts contributed to the war effort. Yet this is only part of the story. Nightingale's commitment to help wounded soldiers and her brave travels to the battlefield often overshadow her work in public health and epidemiology.

Compared with Nightingale's work as a nurse, which is the subject of endless biographies and articles, her work as a public health authority or epidemiologist is barely recognized. Only a few scholars have documented her work as a statistician. I am building on this scholarship by arguing that she was an epidemiologist, based on her efforts in disease prevention, sanitation, theories of disease transmission, and the development of civil engineering practices, particularly her efforts to develop blueprints for hospitals.[22]

Nightingale was interested in using her observations to learn what caused disease to spread and how to prevent it. While doctors like Gavin

Milroy and James McWilliam emerged as leading epidemiologists in the first part of the nineteenth century, Nightingale became a key theorist in the study of disease in the second half of the century. Nightingale demonstrated that nurses offered keen insight in investigating disease. A "careful observing nurse," she stated, can offer a "more important class of data" than a doctor who only periodically visits the sick.[23]

When she returned to London in 1856 a few months after the Crimean War ended, Nightingale compiled the reports and observations that she had collected during the war. The secretary of war had required her to examine the state of the hospitals, particularly their "defects," and to offer an assessment of the health conditions of the army. Her findings were published in 1858 in a book of over eight hundred pages entitled *Notes on Matters Affecting the Health, Efficiency, and Hospital Administration of the British Army.*[24]

But perhaps the most significant aspect of her career was the transformation in her ideas about the cause of high mortality in military hospitals. After she arrived in Scutari, the mortality rate increased instead of decreasing. During her first winter (1854–1855), an estimated 4,077 soldiers died from cholera, typhoid, and typhus. She blamed the high death toll on poor nutrition and the failure to deliver supplies. In February 1855, Parliament appointed a sanitary commission to investigate the health conditions in Scutari. Gavin Milroy, whose study of cholera in Jamaica was discussed in Chapter 4, joined the commission to take the place of Hector Gavin in July 1855. The commission pointed to defective sewers and unventilated buildings as the cause for the high mortality rate.[25] This revelation changed how Nightingale understood the spread of infectious disease and inspired her to pay closer attention to the sanitary conditions in the army hospitals. She went on to become a leading investigator and tireless advocate for improved sanitary conditions.

As Nightingale explained in an unsigned article in the *Leeds Mercury,* the report of the royal commission had shown that "there is, in England, no general hospital system capable of being made the basis of general hospital administration during war, and that to the want of such a system at home were to be attributed most of the disastrous results which rendered the hospitals at Scutari little better than pest-houses." The commission had advised that "to avoid similar disasters in future," general hospitals should be established in England. Parliament had received plans to construct a thousand-bed hospital in Southampton, but Nightingale criticized the plans, claiming that they could not be used as a model for an army hospital without risking "another great catastrophe like that of Scutari." The royal commission, she explained, "has clearly shown that such a hospital should consist of separate buildings or partitions, having wards with windows on opposite sides, with the means of thorough ventilation by the windows; that these buildings should be connected together by corridors for the purpose of administration and of exercise . . . that the wards should be of sufficient size for the purposes of clinical instruction and economical administration."[26] In the same way that slavery and imperialism produced new built environments that revealed how disease spread in large populations of people due to changes in social arrangements, the Crimean War enabled Nightingale to better understand how hospitals should be designed.

Her indictment of hospitals captured the attention of Queen Victoria and her husband, Prince Albert, who served as a loyal patron to the Statistical Society of London, which was founded by his tutor, Adolphe Quetelet, in 1834.[27] When Nightingale returned to England in August 1856, the queen summoned her to meet with her and Prince Albert at Balmoral Castle, the Scottish home of the royal family. Nightingale gained their support to establish a royal commission to investigate the health of the army. She then asked William Farr to join her. Farr was a pioneering statistician who was in charge of vital statistics for the General Record Office. He helped analyze the data she had collected about mortality rates in the war.[28]

Working together, Nightingale and Farr discovered that more soldiers died from disease than from combat. For every soldier who died from battlefield wounds, they found, seven soldiers died from preventable disease in the camps.[29] In addition to statistics on mortality, Nightingale published extensive reports that detailed the unsanitary conditions in the hospitals. She also offered recommendations for how the army could prevent the spread of illness. In her book *Notes on Hospitals,* she began by remarking on the case of a person with "a slight fever" who was admitted to a hospital and recovered from the fever after a few days, "and yet the patient, from the foul state of the wards, [was] not restored to health at the end of eight weeks."[30]

Initially she held on to her first report, later published as *Notes on Matters Affecting the Health, Efficiency, and Hospital Administration of the British Army,* which was intended to be a confidential correspondence between the war office and the army medical department, until the royal commission had been officially formed. She approached Lord Panmure to establish the commission in November 1856, but it was not formed until May 1857. Because of this delay, she paid for the printing of the report and circulated it among English politicians and other officials. The report was well received. One reader noted, "I regard it as a gift to the Army, and to the country altogether priceless."[31]

In the report, finally published in 1858, she explained that the unhealthy conditions in hospitals were not limited to the Crimean War, and traced the history of the British Army's sanitation challenges. She noted that typhus had plagued military hospitals during an English campaign in Portugal in 1811–1814. By offering this example, and others like it, she hoped to develop a larger historical context to expose how unsanitary conditions led to the spread of disease. She further hoped that her analysis would prevent "similar disasters in future" and compel the army to create sanitation policies.[32] Her primary concerns, sanitation and the development of preventive methods, are the major tenets of modern public health.

While in Scutari, Nightingale developed a system of recordkeeping that tracked a variety of factors at the Barrack Hospital and the nearby General Hospital. She took notes on everything from cleanliness to the quantity of supplies to diet to the placement of latrines and graveyards. She also carefully examined the physical space. She took careful note of the size of the wards, the condition of the roof, and the quality, size, and placement of the windows.[33] In her book on the health of the British army, like Thomas Trotter and others who wrote about the importance of fresh air, she pointed to the problem of improper ventilation, and she devoted an entire section to "bad ventilation." She quoted the report of the sanitary commission, which remarked on the "defective state of the ventilation" in the Barrack Hospital. There were only "a few small openings here and there," so that there was no way for the "hot and foul" air to escape. As an adherent of the miasma theory, she believed that diseases were spread through the air and advocated for ventilation to release the "foul air" from hospitals.

In addition to inadequate ventilation, Nightingale pointed to poor drainage and badly designed sewers and plumbing. In her testimony to the royal commission, Nightingale reported on the filthy conditions she found in the Barrack Hospital when she arrived. "The state of the privies . . . for several months, more than an inch deep in filth, is too horrible to describe." She observed six dead dogs under one of the windows, and a dead horse lay in the aqueduct for weeks. The drinking water was dirty; once she saw used hospital uniforms in the water tank. Rats and insects abounded, and "the walls and ceilings were saturated with organic matter."[34]

In the conclusion to her report on the health of the British Army, she explained, "We have much more information on the sanitary history of the Crimean campaign than we have upon any other, but because it is a complete exam (history does not afford its equal) of an army, after failing to the lowest ebb of disease and disaster from neglects committed, rising again to the highest state of health and efficiency from remedies applied. It is the whole experiment on a colossal scale." She pointed out that during the first seven months of the Crimean campaign, mortality exceeded that of the plague of 1665 as well as that of recent cholera epidemics. But during the last six months of the war, after sanitary reforms had been made, "we had . . . a mortality among our sick little more than that among our healthy Guards at home."[35]

Using mortality data that she had collected during the war, along with domestic mortality statistics, Nightingale showed that between 1839 and 1853, mortality among soldiers was much higher than among civilian men: "of 10,000 soldiers [at the age of 20], 7,077 live to the age of 39, out of whom 135 die in the next year of age; whereas out of 10,000 civilians at the age of 20, 8,253 attain the age of 39, and of those 106 die in the year of age following."[36] Nearly all mortality among soldiers was the result of disease; "actual losses in battle form a very small part of the calamities of a long war."[37] Nightingale classified the causes of death as "zymotic diseases" (which in the nineteenth century referred to infectious diseases such as fevers, measles, and cholera), "chest and tubercular diseases," and "all other diseases (including violent deaths)." Nightingale was critical of the army's classification system for diseases. At the bottom of a chart, she notes, "Bronchitis and influenza have no place in the Army nomenclature. The chronic catarrh of the Army Returns is believed to be really phthisis, in the great majority of cases; acute catarrh comprehends both epidemic catarrh, or influenza and bronchitis."[38]

Nightingale presented statistics using charts, tables, and diagrams, which were just beginning to appear in research reports, to make it easier for readers to visualize the comparison she was making.[39] She developed a new kind of graphic, called a "rose chart," also known as a coxcomb chart or polar area diagram, to present mortality data from the Crimean War. Each chart, which is laid out like a pie, shows data from one year, with the slices representing months. Each slice is divided into colored segments whose area is proportional to the number of deaths. One segment is for deaths from wounds, a second for "preventable or mitigable zymotic diseases," and a third for all other causes. A quick glance at the charts of deaths from April 1854 to March 1855 and April 1855 to March 1856 is enough to show that many more deaths were caused by disease than by combat, and that overall mortality decreased in the second year.[40]

To further make visible the dangers of unsanitary hospitals, Nightingale gathered mortality data for matrons, nuns, and nurses working in fifteen London hospitals who died of the "zymotic diseases" of fever and cholera.[41] She presented tables, which she notes William Farr compiled for her, showing that the mortality rate of the nursing staff was much higher than that of the female population in London; in addition, women working in hospitals were more likely to die of zymotic diseases than were other women. She used these figures to argue for the "very great importance" of hygiene in hospitals. "The loss of a well-trained nurse by preventible [sic] disease," she wrote, "is a greater loss than is that of a good soldier from the same cause. Money cannot replace either, but a good nurse is more difficult to find than a good soldier."[42]

In her book *Notes on Hospitals,* she retold the story of the British prisoners of war who died in a crowded jail cell in India in 1756: "Shut up 150 people in a Black hole of Calcutta, and in twenty-four hours an infection is produced so intense that it will, in that time, have destroyed

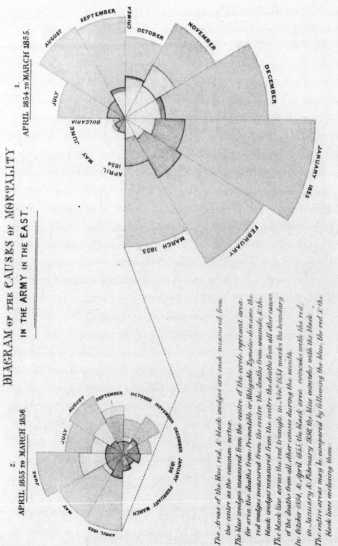

DIAGRAM OF THE CAUSES OF MORTALITY
IN THE ARMY IN THE EAST.

1.
APRIL 1854 TO MARCH 1855.

2.
APRIL 1855 TO MARCH 1856

The Areas of the blue, red, & black wedges are each measured from
the centre as the common vertex.

The blue wedges measured from the centre of the circle represent area
for area the deaths from Preventable or Mitigable Zymotic diseases; the
red wedges measured from the centre the deaths from wounds; & the
black wedges measured from the centre the deaths from all other causes.

The black line across the red triangle in Nov.r 1854 marks the boundary
of the deaths from all other causes during the month.

In October 1854, & April 1855, the black area coincides with the red,
in January & February 1856 the blue coincides with the black.

The entire areas may be compared by following the blue, the red, & the
black lines enclosing them.

Florence Nightingale's
rose chart showing
causes of mortality in the
British Army during the
Crimean War, April
1854–March 1855 (*right*)
and April 1855–March
1856 (*left*). The area of
each wedge is propor-
tional to the number of
deaths for that month.
The pale wedges (red in
the original) indicate
deaths from wounds;
light gray (blue in the
original), deaths from
disease; dark gray
(black in the original),
all other causes.
(Wellcome Collection)

nearly the whole of the inmates."[43] Nightingale's reference to the case is evidence for its status as the prototypical illustration of the need for ventilation. And the fact that it took place in India shows how British medical authorities used information from around the empire.

As a result of her work with large numbers of patients in the Crimean War, Nightingale framed her analysis like an epidemiologist, in terms of populations. She focused on how disease spread within a group. She devoted her energies not to changing bedpans or dressing wounds but to studying the structure of hospitals, analyzing statistics, and figuring out how to increase ventilation.

The war provided her the opportunity to compare mortality rates in varied settings: crowded hospitals, shabby tents, and wooden huts. It also underscored to her the importance of preventive measures, which constitutes one of the major tenets of modern epidemiology. By publishing her observations, her insights, and guidelines for hospitals to follow, she hoped to provide a set of rules and guidelines for physicians to follow to prevent the spread of disease. While efforts to ensure proper hygiene as a way to guard against illness can be traced to Mesopotamian civilization and Sanskrit writings from 2000 BCE, Nightingale's warnings, in particular, and sanitary reform, more generally, sparked a critical turning point in the middle of the nineteenth century that gave rise to preventive medicine.[44] This transformed military medicine from an enterprise that largely focused on treatment and surgery to one that began to engage epidemiological questions and issues.

By drawing on her research, Nightingale challenged some of the leading debates of the period. In particular, she discussed the debate concerning "the communication of 'infectious' disease, both in civil and military hospitals, from patients' linen to washerwomen." She pointedly asked, "Have those who put forward this doctrine of inevitable 'infection' among washerwomen ever examined the process of washing, the appliances by which it is done, and the place where the women wash?" "If they will do so," she continued, "they will find a small, dark, wet, unventilated, and overcrowded little room or shed. . . . Is it surprising that the linen is badly washed, that it is imperfectly dried, and that the washerwomen are poisoned by inhaling organic matter and foul air?" She further asserted that it would be helpful to place a moratorium on the issue of "infection" and instead devote efforts to reform "washing establishments and convert them into proper laundries." She called for "sufficient area and cubic space for each washer, with abundance of water," with proper drainage, ventilation, and separate rooms for drying and ironing. Improving sanitary conditions, she contended, would prevent washerwomen from "'catching' fever."[45] She framed the need for an "abundant supply of pure water" in terms of sanitation. Like Gavin Milroy and other sanitary reformers, she emphasized the importance of proper drainage. She preferred to provide persuasive evidence of specific conditions in known places in order to explain the need for proper sanitation than to rely on a theory to make the point. She traced disease to crowded conditions and called for preventive protocols to stop the spread of disease but believed that "inhaling organic matter and foul air" caused sickness.

Although Nightingale devoted her life to writing about the transmission of disease, her ability to observe the actual conditions within a large number of hospitals ended after she returned from Crimea. While stationed at the Barrack Hospital during the Crimean War, she became deathly ill. While it remains unclear what actually ailed her, most scholars believe that she had contracted brucellosis, a highly contagious bacterial infectious disease transmitted from animals to humans often from exposure to undercooked meat or contaminated animal products like raw milk.[46] She described being "a prisoner to my bed" and was confined to her room for the rest of her life.[47] Despite these obstacles, she remained committed to the study of public health and disease transmission.

While physicians had long been pointing out the dangers of overcrowding in prisons and on slave ships, the Crimean War served as compelling evidence about the dangers of crowded spaces, especially among US physicians during the Civil War.[48] Based on her observations that overcrowding and lack of fresh air in hospitals led to disease, Nightingale argued for proper ventilation.[49] Her campaign reveals how scientific ideas often need to be reiterated with urgent examples in order to gain widespread public acceptance.

✦   ✦   ✦

Shortly after Nightingale's return to England in 1856 came the Indian rebellion of 1857–1858, which shifted British government, military, and public attention to South Asia. British colonialism in India had begun when the East India Company was formed in 1600. Established as a network of London merchants who imported spices from India to England, the East India Company grew into a powerful military and political organization that exerted imperial control over vast regions of the subcontinent. When a war between France and England spilled into India in the 1740s, the company expanded its operations and became a military force by recruiting local Indians to join its new army. With military power, the East India Company dominated European trading companies in India and overthrew local leaders. By 1756, the company had recruited 20,000 Indian men as soldiers into its army, who became known as sepoys. By 1803, the size of the army had multiplied by ten, to 260,000.[50] Throughout the first half of the nineteenth century the number of sepoys increased to an estimated 311,000, compared with roughly 40,000 Englishmen serving in India.[51] In 1857, due to a range of grievances, including the unsanitary conditions of the barracks for Indian soldiers, sepoys from all over India rose up in rebellion against British forces. The British ultimately defeated the sepoys. The immediate result was that the British government took control of India by an act of Parliament in 1858, transferring power from the East India Company to Queen Victoria.

Nightingale decided to take advantage of the expansive bureaucracy that British imperialism had created in India. She embraced Edwin Chadwick's idea of creating a royal commission to investigate sanitation conditions of the army in India and to offer recommendations to the government. Sidney Herbert, who had sent her to Scutari, worked with her to put together the Royal Commission on the Sanitary State of the Army in India (1859–1863), which included military and civil physicians, sanitarians, the queen's doctor, and a lawyer. Nightingale then created an inner circle that included William Farr, who had worked with John Snow and served as the statistician for the General Register Office, the sanitarian John Sutherland, and Herbert, who served as chair of the commission.[52]

Nightingale believed that sanitary laws should be "as much part of the future régime of India as the holding of military positions or as civil government itself." She noted that a British officer had appointed a sanitary inspector "to attend to the cleansing of a captured city and to the burial of some thousand dead bodies of men, horses, asses, bullocks, camels and elephants, which were poisoning the air," but that the government in Bombay had not sanctioned the cleanup because there was "no precedent." She did not expect that British intervention would be a permanent force in maintaining proper sanitation, and hoped that the local Bombay government would eventually take over these efforts. Despite this recognition, Nightingale, like many other people in power throughout the nineteenth century and beyond, recognized the power in using science to subjugate a population. She asserted that British authorities could use "hygiene as the handmaid of civilization."[53]

Nightingale and other British authorities indicted health conditions in India based on Victorian English standards.[54] Leaving aside the problems of sanitary conditions in England, which she and others aimed to reform, Nightingale and the royal commission overlooked the health and healing practices that local people adhered to; they ignored their understandings of disease and treatment; they failed to recognize their local customs, culture, and values. Instead, they defined India as a region in need of British intervention.

While science and medicine certainly functioned as tools that facilitated imperialism, imperialism also contributed to the development of science, in the same way that it contributed to the debate on contagion and quarantine in Cape Verde. Imperialism produced a huge bureaucracy that provided Nightingale with massive quantities of records to study for information on sanitation and disease transmission. It allowed her to put into practice the principles that underpinned the burgeoning field of applied statistics. Nightingale recognized the value of India as a source of her intellectual work and exclaimed in 1879, over twenty years after she spearheaded the commission, "my interest in India can never abate."[55]

Due to her illness, Nightingale never went to India, but she had access to numerous reports and committee minutes and corresponded with many administrators.[56] As the field of public health developed, practitioners began to rely more heavily on reports generated by the bureaucracy, which provided a detailed overview of the region. During Nightingale's collaboration with Farr during the Crimean War, she had already analyzed statistical data about regions that she never visited. By studying sanitation from her bed, not on the frontlines in India, she demonstrated that scientific knowledge production did not require her to be in India. The emerging field of epidemiology depended less on immersion in a particular region and more on analyzing data and making recommendations—though she often wished she could travel to South Asia. In many of her letters, she exclaimed that she wished she could go to India.[57] Her intellectual work, nonetheless, remained important to the development of epidemiology despite the fact that she never set foot in India.

Nightingale did not need to go to India in order to study it; she instead relied on new epidemiological methods. The medical bureaucracy had expanded and had officially become invested in documenting health conditions. The Royal Commission on the Sanitary State of the Army in India dispatched physicians to provide medical services to the army and to monitor the health conditions. Nightingale had persuaded many medical and government officials to take health seriously, and as a result, many sanitary measures were adopted. The establishment of the commission led to the development of more detailed reports—official government records and manuals, minutes, and memoranda from meetings—and an increase in medical officers charged with observing and documenting health conditions in India. Because Nightingale had gained a reputation as a leading medical authority, she also had direct contact with many military officials and medical experts who had served in India; when they returned to London, they shared with her their observations, official correspondence, and other materials that informed her understanding of health in the region.[58] Imperialism in India produced such an expansive bureaucracy that Nightingale could become an expert without leaving England. Data could be generated about a region and then processed someplace else. For example, in Nightingale's 1863 report on India, she offers precise descriptions about a place she had never visited. "Allahabad, one of our largest and most important stations in one of the worst positions, as if that position were not unhealthy enough by itself, trusts to nature again, has no drainage nor sewerage, and leaves its surface water to 'evaporate,' 'percolate,' and 'run off.'"[59]

Nightingale drafted questionnaires that the royal commission sent to 150 to 200 stations throughout India about health conditions. The completed replies were called stational returns. Nightingale analyzed this information with the help of her "cabinet"—Sutherland, Herbert, and especially Farr, who focused on the statistical information relating to mortality and birth rates and the spread of infectious disease. The commission also interviewed some of the military officials who submitted stational returns. The commission's final report, mostly written by Nightingale, was published in 1864.[60]

Imperialism had disembodied knowledge production; the person who collected the information was no longer the one who analyzed it. Certainly, gathering information relies on analytical choices, but the expansion of bureaucracy allowed for the production and circulation of documents that enabled experts to examine health conditions without having to be on location. This feature of imperialism also had an effect on how studies of disease transmission were conducted in other regions. In London, New York, Paris, and many other major cities in the nineteenth century, doctors formed large associations to study epidemics; in doing this, they, too, drew on reports furnished through bureaucratic channels.[61] Imperialism had popularized the practice of using statistical knowledge and narrative reports to study health.

More specifically, statistics served as a tool of empire; it facilitated the aims of imperialism by offering a narrative of the people and place that seemed, from the vantage point of British authorities in England, unwieldy. British authorities had depended on statistics to document the number of people in a region, to track mortality and birth rates, and to quantify differences between European troops and sepoys.[62] Government, medical, and military officials came to understand India through statistical descriptions. Statistics provided British authorities with a set of analytical points that they used to advance their control over the country.[63] Knowing the number of Indian people in a region, for example, made it possible to determine the number of soldiers needed to control them, and statistical data could also alert authorities to the dangers of a particular region where British soldiers or local people were succumbing to epidemics. Studies of India exemplified the value of statistical knowledge as a metric in the emerging field of epidemiology.

The reliance on statistics to evaluate the spread of infectious disease did not simply result from Farr's collection of statistics during the cholera epidemic in Soho in England; it can also be traced to Nightingale. In the mid-nineteenth century, statistics remained a virtually new field of inquiry in the emerging social sciences.[64] Nightingale quickly emerged as a leading figure. The *Journal of the Royal Statistical Society,* for example, published statistics from British hospitals based on methods she had recommended.[65]

In 1858, Nightingale became the first woman elected to the Royal Statistical Society. The 1860 International Statistical Congress applauded her research, stating, "Miss Nightingale's Scheme for Uniform Hospital Statistics should be conveyed to all governments represented."[66] Nightingale had written to Lord Shaftesbury to encourage governments at the 1860 meeting in London to publish their extensive use of their statistics. Various hospitals in England adopted Nightingale's methods in 1862 and published their findings.

Nightingale's ascent as an authoritative statistician continued with her dedicated analysis of health in India. Her bedroom was carpeted with reports from India, and she was spending enormous amounts of time interviewing medical experts, military officials, and other visitors about health in India. Her work with the royal commission made her an authority on statistics and helped shape the field. In the early twentieth century, the statistician Edwin Kopf published an article in the *Journal of the American Statistical Association* that praised her statistical work and put her contributions next to those of Quetelet and Farr.[67]

Nightingale called for the 1861 census of the United Kingdom to be revamped to include an enumeration of the number of the "sick and infirm." "We should have a return of the whole sick and diseases in the United Kingdom for one spring day, which would give a good average idea of the sanitary state of all classes of the population," she wrote in a letter to William Farr.[68] She also urged collection of information about housing. By using the census to collect information about where sick people resided, she hoped to collect more specific information than just the number of people who died each year in the United Kingdom. She hoped to map the relationship between housing conditions and the spread of disease. Her request made it to Parliament in 1861 as part of a Census Bill, but ultimately her suggestion was denied.[69] Despite this failure, the United States and several other countries later included sickness and housing questions on their census forms.[70]

÷ ÷ ÷

Nightingale continued to investigate health conditions in India throughout the rest of her life, but at the same time, science was taking a radical shift. In the 1870s and 1880s, the germ theory of disease was developed, primarily by the French scientist Louis Pasteur and the German physician Robert Koch. When cholera broke out in Egypt in 1883, the German government appointed Koch to lead a commission to study it. After the epidemic subsided in Egypt, Koch asked the German government for authorization to travel to Calcutta, India, where the disease was continuing to spread. In Calcutta, he conducted hundreds of autopsies and discovered the presence of bacteria in a local water "tank," an open pool of water where people bathed, washed, and collected water for drinking, from which seventeen cases of cholera could be traced. He then isolated the bacterium and was able to culture it. Using a microscope, he described the cholera bacillus as "a little bent, like a comma," unlike other bacilli, which have a rod-like shape. He found the bacillus in the stools of those infected with cholera as well as on soiled linen, but not in patients with other forms of diarrhea. His discovery advanced John Snow's waterborne theory by pinpointing the microscopic element in the water that caused cholera.[71]

Nightingale initially dismissed Koch's theory and continued to advocate for sanitary reform. Her colleague Dr. John Sutherland, however, purchased a "beautiful Vienna microscope" in order to observe the bacillus. After this, Nightingale eventually and begrudgingly accepted the theory, but with reservations.[72] Like many others in the medical profession, Nightingale refused to accept germ theory as the sole explanation of disease transmission, even when many epidemiologists at the time accepted it, because she believed that it overemphasized the agency and power of germs over the problem of unhealthy sanitary conditions.

Germ theory also reinvigorated the debates about contagion and quarantine that had polarized the medical profession, government authorities, merchants, and the military in the first half of the nineteenth century. Contagion theory blamed a sick person for spreading an infectious disease, which led to quarantine restrictions and isolation of the infected person from the rest of the population. Germ theory rested on a similar premise: the germ, like the infected person, was the source of disease spread. According to Nightingale, supporters of germ theory believed that isolating or eradicating the germ would prevent epidemics. Nightingale and others put forward a more nuanced argument, contending that the physical environment must be taken into consideration as a factor in the spread of infectious disease.[73] She pointed to how filth, foul water, inadequate ventilation, and other unsanitary conditions caused the outbreak of disease.

Nightingale argued that the focus on germs obscured environmental factors.[74] In a letter to Thomas Gillham Hewlett, sanitary commissioner for Bombay, in 1883, Nightingale wrote, "Our whole Indian experience tends to, nay actually proves, that cholera is not communicable from person to person, that it is a local disease, depending on pollution of buildings, earth, air and water, that quarantines, cordons, medical inspection and the like are all fatal aggravations of the disease." She argued that the only way to prevent the spread of cholera was to "remove healthy troops, healthy people, from the locality, to put earth, air, and water and buildings into a healthy state by scavenging, limewashing and all sanitary work."[75]

Nightingale then referred to the situation in Egypt to further illustrate the urgency of sanitation. If British authorities in Egypt had undertaken the same sanitary measures as Hewlett had put in place in Bombay, the cholera epidemic "would never have come" or would have been "a slight epidemic." Instead, she wrote, in both Egypt and Europe, "It is this doctrine of 'germs' which has 'poisoned' us."[76]

Nightingale came to believe in germs by the late 1880s, but she questioned their origin. She contended that filth produced germs and that sanitary measures were thus needed to prevent their spread. Koch's research in India suggested the reverse. He showed that the human body contained cholera bacilli, which could be found in stool. Due to faulty sewers and poor sanitation, stool could spill into the waterways, causing the bacilli to spread. In essence, Koch and Nightingale disagreed about what came first. Their arguments were not mutually exclusive: maintaining proper sanitation and monitoring sewers, drains, and the water supply would prevent the spread of the bacteria. But at the time, from Nightingale's perspective, these two theories belonged to two separate camps: doctors, who believed in addressing germs, versus sanitarians, who believed in addressing the physical environment and public health. Nightingale worried that doctors would be empowered to make decisions about public health, and sanitation efforts would be defunded. In an 1886 letter to Lord Dufferin, viceroy of India, Nightingale submitted that "sanitation is a specialty; ordinary medical men are contagionists and would advise quarantine and such things. And not one has studied the question of the sanitary construction of the buildings, etc."[77]

Nightingale further explained to Dufferin that science was not just a matter of theories; it was also a matter of practice. She pointed to her work as a statistician. "Disease and death reporting have done their work" over the past twenty-three years, she wrote. She noted that death rates had decreased among British soldiers and sepoys and that mortality rates in jails had also declined. She contended that death rates in cities had "shown some improvement" but that the "results show mainly what *might* be done" and that city governments wanted to know how to execute measures to improve health conditions. For Nightingale, sanitation was a matter of practice, whereas germs were a matter of theory. She continued to advocate for measures to be adopted on the ground to maintain healthy conditions.[78]

In general, Nightingale viewed Indian health not as "a natural production of India" but as a result of "rational management." Like Trotter and Milroy, she challenged the long-held British belief that tropical climate and foreign topography caused the spread of infectious disease. Nightingale posited a bold, rather sophisticated argument that health conditions result from the built environment, from human decisions rather than from the natural landscape or climate. While at times she propagated negative, racist stereotypes about India as filthy and the people as savage, she saw these conditions less as a result of innate inferiority than as a product of circumstances that could be altered.[79]

Nightingale's arguments, which she rehearsed from her Victorian bedroom in London, only became possible because of imperialism. Imperialism had provided her with the data that illustrated the need for proper sanitation. She had spent years studying reports on the buildings in India, the waterways, the sewers and drainage, as well as statistics on birth, death, and disease rates. British imperialism in India enabled her to analyze these phenomena, which is why she refused to deny the importance of sanitation. Imperialism in India had allowed the gathering of statistics that gave her further evidence that disease thrived in unsanitary environments. These results inspired her to continue to challenge those who believed in germ theory as the sole explanation for the spread of disease. This then led to important, healthy debates that influenced the development of epidemiology, public health, sociomedical science, medical anthropology, and many other subfields and discourses for decades. As a result, imperialism is at the root of many genealogies of modern science.[80]

Indeed, Koch's discovery of the cholera bacillus was made possible by imperialism as well. It transpired during the so-called scramble for Africa, when British, French, and German powers were in a fierce race to colonize Africa and expand their empires.[81] Imperialism empowered Koch to study the water in Calcutta and to do autopsies; it licensed him to track the presence of the bacteria in the dead and in the soiled linen of the living.

Imperialism provided Koch and his team of experts with the evidence to track the effects of cholera in the village near the polluted water source, which provided proof about the cause of the epidemic. And yet, very little about Koch's foray in India and the people whom he studied has been central to the narratives that chart his rise to fame. Instead, the focus remains on his ideas—not the people. The cholera epidemic in Egypt exploded at the time that European powers were attempting to gain control of the Suez Canal. The expansion of European imperialism in the region led to competition among England, France, and Germany in order to understand how infectious disease spread. Koch's discovery gave the German Empire a lead, which then caused the British to refute Koch's findings in order to advance their imperial authority.[82] British authorities rejected Koch's theory based on several considerations, including the fact that a number of people who drank water from the polluted water tank that Koch had studied did not contract cholera.[83] This debate would not have been possible without the people whom both empires counted in order to substantiate their theories.

This debate cast European imperial authorities and their dispatched medical teams as the leading protagonists in this narrative, and treated imperialism as a race among European nations to enlarge their empires without any reference to the actual people who suffered and died from cholera. Imperialism subjugated a population of people and turned them into objects of study that European powers used to investigate the cause of the cholera epidemic. The story of how subjugation led to scientific theories eventually evaporated from the historical record, if it was even noted by those in Europe in the first place. A new narrative solely about science emerged, with little to no reference to autopsies, the sick people by the water tank, the people living near the polluted sewers who drank from the contaminated tanks. No references surfaced of the names of the people whose lives were reduced to a number. Imperialism turned medical thinkers, like Nightingale and Koch, into leading figures, while no one paused to consider how they managed to gather their data or what circumstances made the evidence readily available.[84]

÷   ÷   ÷

While Koch has been credited for his accomplishments, Nightingale has been regarded by the public more as the "lady with the lamp" during the Crimean War than as an epidemiologist. She spent only a short time as a nurse offering care and comfort to patients in hospitals, however; most of her career was spent thinking, writing, debating, analyzing, collaborating, and publishing reports on disease transmission. The reason for this elision rests in part on the fact that advances in scientific knowledge are usually traced to white men in Europe, not to women working in makeshift hospitals in Scutari and or in their bedrooms analyzing British troops and local communities in India. The frameworks of knowledge production have obscured Nightingale's work and narrowly cast her as a nurse instead of an epidemiologist. This is partially due to the fact that Nightingale is often studied in isolation as the subject of biographies or articles. While her work increasingly appears in studies about the origin of public health, it remains divorced from that of other medical professionals who studied the spread of infectious disease at the time.[85] Like Thomas Trotter, Gavin Milroy, and others, Nightingale traveled abroad, studied medical conditions throughout the British empire, investigated the cause of illness, established protocols, and published her findings. Her vast archive of publications makes a powerful case for her role as an epidemiologist.

That said, Nightingale's work as a thinker was not entirely divorced from her work as a nurse. The combination of war and imperialism shifted her focus from the care of individual patients to public health. She drew on her experience at the hospitals in Scutari to advocate for proper ventilation, rules prohibiting overcrowding, and cleanliness, and she provided equations to use for calculating the amount of space needed for hospital corridors and patient rooms. Even when she was bedridden and unable to travel, she continued to investigate health conditions by examining the environmental and physical conditions of particular locations. Because of her focus on drains, sewers, and other environmental factors, she resisted accepting germ theory, which seemed to her to undermine the gritty realities that she observed in Scutari and read about in India, Russia, and other parts of the world. She remained intellectually committed to miasma theory and questioned the existence of microbes, but she did play an important role in establishing a set of practices that contributed to the field of epidemiology. Like Gavin Milroy in Jamaica and James McWilliam in Cape Verde, her methods and subjects of analysis mattered more than her position on any given theory. Her emphasis on environmental conditions advanced public health; her careful analysis of mortality, birth, and sickness rates helped to formalize statistical analysis as a key hallmark of epidemiology; her reports and correspondence advanced the role of bureaucracy in tracking and studying epidemics.

Consequently, Nightingale's contributions to epidemiology resulted from her efforts to elevate ideas about disease prevention to a theory that transcended the specifics of a particular place or people. The field of epidemiology depended, in part, on evidence from India and the Crimean War to solidify its claims. While Thomas Trotter, Arthur Holroyd, and others from the early nineteenth century had advanced new ways of thinking about the cause, spread, and prevention of disease, their ideas had not been formally codified as recognizable epidemiological practices. Their studies functioned as case studies that exposed the problem of crowded spaces, raised awareness about the need for local governments to maintain sanitary regulations, and created methods for tracking the spread of disease that provided the foundation for epidemiology. Milroy's work, however, signaled a shift. He was known as an epidemiologist. Nightingale, while never referring to herself as an epidemiologist or being part of the all-male Epidemiological Society of London, nonetheless established a number of practices, particularly the use of statistics to track epidemics, that shaped contemporary epidemiology and public health. Though she was not recognized as an epidemiologist at the time, the historian has the ability to interpret the evidence, place her medical insights in context, and assess the significance of her contributions to the field.

During the mid-nineteenth century, Florence Nightingale's ideas gained prominence throughout the world. Doctors, reformers, and others recognized the value of her theories and applied them to other peoples and places. When war broke out in the United States between the North and South in 1861, military, civilian, and medical officials turned to Nightingale's theories and put them into practice.

CHAPTER 6

From Benevolence to Bigotry

The US Sanitary Commission's Conflicted Mission

WHEN THE US CIVIL WAR BEGAN IN 1861, President Abraham Lincoln called on Americans to support the war effort. Many middle- and upper-class white women saw the war as an opportunity to expand their efforts. They had read about Florence Nightingale's heroic efforts and became disciples of her precepts, viewing health and hygiene as a new frontier to benevolence. While Nightingale's response to the Crimean War advanced medicine, the work of women reformers during the Civil War unwittingly created a Trojan horse that enabled physicians to codify racist understandings of Black people and in so doing reverse the advances of epidemiology.[1] US doctors, unlike their British counterparts, formally classified racial difference as a key hallmark in examining the spread of infectious disease, which had the disastrous effect of solidifying racial identity as a central component in epidemiology and public health.[2]

Elizabeth Blackwell emerged as a leading figure during this period. In 1849, she became the first woman in the United States to earn a medical degree. Like Nightingale, she believed that disease resulted from poor social conditions and an unhealthy environment. In fact, her graduate thesis identified typhus as the cause of illness among many Irish immigrants to the United States. Born in England, Blackwell had met Nightingale in 1850. The two women initially shared similar understandings about the importance of women in medicine, but this later changed. Nightingale wanted Blackwell to train nurses, but Blackwell believed that nursing would solidify women's inferior status within the broader field of medicine. Despite this difference, in her practice in New York City, Blackwell continued to follow many of Nightingale's key insights about sanitation as the most effective measure to prevent the outbreak of disease.[3]

When the Civil War began, Blackwell took her cue from Nightingale, believing the war would provide an opportunity to usher women into the medical profession. She initially attempted to centralize the various reform efforts that had already sprung up, merging them together with the existing societies that predated the war. She organized a meeting in late April 1861 that brought women reformers together at Cooper Union. There, they established the Woman's Central Association of Relief (WCAR), which had three major objectives: to work with the army's medical department, to train nurses, and to unify women's relief work.[4] Blackwell and her female colleagues realized that they needed a man to be the face of their organization if they were to be recognized as a legitimate. Henry Bellows, a minister, stepped forward to serve as the leader. He was to go to Washington, DC, to meet with the president to share with him their plans.

Meanwhile, Dorothea Dix, who had already gained a reputation as an advocate for intellectually disabled people, had begun her own campaign. Like Blackwell, she followed Nightingale's principles, and she had even tried to meet Nightingale. In 1855–1856, during the height of the Crimean War, Dix traveled to Europe and toured asylums in order to gain insight about practices and treatment. She also went to Scutari, where she just missed Nightingale but was able to witness firsthand her efforts within the hospital wards.[5] During her trip, she also likely learned how nurses could use political strategy to navigate their way through the military bureaucracy; when the Civil War broke out she went directly to the secretary of war, Simon Cameron, in the same way that Nightingale had approached Sidney Herbert for assistance.

Bellows eventually teamed up with Dix because he thought she was better suited than Blackwell to support his campaign for sanitary reform.[6] After leaving New York, Bellows had toured army camps and witnessed the unhealthy conditions, and he began to realize that New York reformers needed to expand the scope and size of their endeavors. Not only would they need to provide clothing, bandages, and other medical supplies to the army, but they would also need to address the sanitary problems that plagued the camps. Army officials initially rejected the proposal for a sanitary commission, believing that the mostly women-based organization would be "obtrusive." But Bellows reassured military authorities, the secretary of war, and President Lincoln that a civilian corps would support the army, serving as a necessary buffer between the army's urgent needs for support and women's determination to get involved. Bellows promised that men would lead the efforts of the commission and that the women would not meddle in the army's operation. Lincoln was persuaded and signed a declaration on June 13, 1861, that approved the creation of the United States Sanitary Commission (USSC).[7]

Because of the bargain that Bellows had struck with the federal government, men led the organization and the women were left behind. They were forced into subordinate positions as assistants to army personnel. Blackwell had hoped that the war would offer an opportunity to train women as physicians, but her vision for the USSC got lost in Bellows's bargain. Blackwell continued to work with the WCAR, but she trained women as nurses, not as doctors.[8]

The First Battle of Bull Run, also known as the Battle of Manassas, on July 21, 1861, revealed to military and government authorities the medical crises that military engagement engenders. Hundreds of wounded men were left on the battlefield or evacuated to makeshift hospitals. Beyond the battlefield casualties, the war also disrupted the environment, which contributed to the spread of infectious disease. Assembling men into regiments created severe population shifts that drained natural resources and limited the water supply; moving troops from one location to the next uprooted ecological systems and damaged the land; establishing camps led to the erosion of agricultural landscapes and the destruction of trees. Multiple health crises resulted from these dramatic transformations, but there was no system in place to address these problems.[9]

Florence Nightingale's work could have warned Americans about the health crises that war produces. Some civilians had listened and slowly adopted her sanitary measures, but many in the military dismissed her ideas and instead privileged strategy over sanitation, politics over public health, and victory over ventilation. At the time of the Civil War, the surgeon general's office served more as a sinecure for retired military officials than as an effective medical corps. As a result, there was no infrastructure to address the dire condition of injured soldiers, the presence of human and animal corpses, and the sanitary problems in army camps. Captain Louis C. Duncan, who published an article in 1917 on the shortcomings of the US Army's medical department at the start of the war, wrote, "The Surgeon General of the Army in 1861 was no doubt a worthy gentleman, [but] he was about as much prepared for war as were the people of San Francisco for an earthquake."[10] Although there were army doctors, there was no hospital system. According to Duncan, the first field hospital was not developed until the Battle of Shiloh in April 1862, when a doctor took over an abandoned regimental camp and converted it into a tent hospital.[11]

The Union Army lost the Battle of Bull Run, which then raised urgent questions about the army's preparation for war. As Duncan explained, it was not just that the army lost the battle but that "some widely prevalent ideas suffered a rude shock that day. One was that a collection of armed men constitutes an army." In addition, regardless of how patriotic or qualified the civilian doctors might have been, they were unprepared for the challenges of wartime medicine. At the time of the battle, there were "no previous plans, no organization, no enlisted personnel, no supplies, no ambulance corps, no field hospitals, no convoys for wounded, no evacuation hospitals." Indeed, "the suffering of the wounded at Bull Run shows the natural result of an inadequate Medical Department." No one in particular was at fault "for this miserable neglect," he claimed; rather, "the era of scientific care of the wounded had not arrived."[12] The Battle of Bull Run signaled to many military officials throughout the Union Army the need for organized medical assistance.

After the Battle of Bull Run, army officials stationed throughout Virginia and Maryland flooded the headquarters of the USSC headquarters, which was temporarily housed in the Treasury Building in Washington, DC, with requests for beds, iron cradles for supporting wounded limbs, tables for writing in bed, and even dominoes to keep patients entertained.[13] The commission responded but also wanted to offer advice on sanitary issues, such as how to design camps to ensure proper ventilation, and how to gain access to clean water.

The USSC helped to advance the field of public health by urging the military to adopt sanitary measures. Charles J. Stillé, a member of the US Sanitary Commission who later wrote its institutional history, explained the influence of the British Sanitary Commission on the USSC and the significance of the Crimean War to the development of sanitary science. At the beginning of the Civil War, he wrote, "the experience of the Crimean war was fresh in the memory of all," and that experience "was a complete chapter by itself on sanitary science."[14] The USSC drew on these ideas; in fact, Stillé knew that British epidemiologist Gavin Milroy went to Crimea as a member of the British Sanitary Commission to respond to the medical crises the war had caused, which further reveals the global network of epidemiologists.[15]

The mere practice of keeping hospitals clean and distributing supplies transformed health conditions within army camps. While it is difficult to determine the effectiveness of the USSC's intervention, its work, nonetheless, emphasized to the public and many in the medical profession that the transmission of disease resulted from conditions in the physical world. Sanitation was, of course, not a magic bullet that could stop the spread of bacteria and viruses (unknown at the time), and its effectiveness is difficult to determine. I am less interested in mortality rates than in how the war stimulated new ways of thinking about the spread of disease and helped to establish preventive measures. Sanitation as a health practice gained momentum because of the war; it drew on transatlantic and global theories and accelerated understandings about disease transmission to a broader public.

In 1861, the USSC issued a fourteen-page document, "Rules for Preserving the Health of the Soldier," that offered tips for disease prevention. Like Nightingale, it emphasized prevention. The pamphlet, which was distributed to Union Army officers throughout the country, offered forty-one concise, clear rules, which ranged from ordering the soldiers not to sleep in their uniforms but rather in "their shirts and drawers," to recommending that soldiers bathe with water at least once a week, to more specific instructions. Bathing once a week represented a new form of cleanliness in the nineteenth century, which differed from the earlier practice of simply changing linens as a way to clean the body. The manual also gave details about grooming hair.[16]

The rules urged military officials to ensure that camps were kept as clean and orderly as possible in order to prevent the spread of infectious disease. "There is no more frequent source of disease, in camp life, than inattention to calls of nature," begins rule 14. It instructs that trenches for defecation should be built "a moderate distance from the camp," with separate trenches for soldiers and officers. Officers should forbid soldiers from defecating elsewhere. The trenches should be visited by a "police party" every day, who would throw in a layer of earth, lime, and "other disinfecting agents." The camp police, continued the rule, should also make sure that offal from slaughtered cattle "is promptly buried at a sufficient distance from camp, and covered by at least four feet of earth."[17]

Other rules within the pamphlet reflect Nightingale's influence, particularly in regard to ventilation and the placement of the tents.[18] Commanding officers might have been more interested in developing strategies and waging battles than in paying attention to where soldiers pitched their tents, but the USSC pamphlet offered clear regulations to ensure that tents were not placed close together and that soldiers were not crowded within them. "Experience has proved," reads rule 18, "that sleeping beneath simple sheds of canvas, or even in the open area, is less dangerous to health than over-crowding in tents."[19] The reference to "experience" in this rule comes directly from the Crimean War.[20] Also drawing from lessons learned during the Crimean War was the USSC's guidance that "camp fires should be allowed whenever admissible" because of their usefulness "for purifying the air, for preventing annoyance from insect," and for providing warmth and drying clothes. As an advocate of miasma theory, Nightingale believed that fireplaces facilitated ventilation and that fires purified the air. Although scientists had not yet discovered that mosquitoes were carriers of malaria, yellow fever, and other diseases, this particular practice had the effect of helping to reduce the spread of mosquito-borne diseases.[21]

Yet, to military personnel, these rules probably appeared as feminizing.[22] Concerns about bathing with water once a week or grooming hair might have been dismissed as trivial. The USSC pamphlet was thin and pocketsized, and it could easily have been disregarded if military commanders did not want to follow the instructions. While there is no archive of data available detailing army officials' rejection of the USSC rules, the large number of soldiers who died from preventable diseases or, to use military parlance, "camp diseases," signals that many officials did ignore these rules. As in the Crimean War, more soldiers died from infectious disease in military camps and hospitals than from military combat.[23] Soldiers coughed to death in their camps or died of dehydration from severe diarrhea.

The war produced a massive bureaucracy that led to the widespread distribution of new theories about sanitation that amplified the development of epidemiology. The USSC helped to make ideas about ventilation, bathing, and the organization of physical spaces available to a larger public.[24]

✦   ✧   ✦

The United States Sanitary Commission changed the face of the Civil War. Although the military at first objected to women being on the battlefield, the USSC, like the British Sanitary Commission during the Crimean War, was eventually allowed to bring women into army camps and onto the battlefields in an official capacity. Before this period, many women followed armies unofficially. Some of these "camp followers" were cooks and aides, some were wives, some were sex workers. The USSC formalized the place of women in the camps, where they carried out a range of duties—dressing wounds, wrapping bandages, cleaning tents, removing dead animals, boiling water, cooking food, pitching tents, and keeping notes of their work. Some of them were lionized for their tireless and devoted service and were represented, like Nightingale, as angels on the battlefield. *The Liberator*, a leading abolitionist newspaper, reported on the courageous work of one woman, whom we know only as Mrs. Reynolds, who cared for the wounded at the Battle of Shiloh and had been made a major in the state militia by the governor of Illinois. The news report was followed by a poem describing her on the battlefield, where cannons roared and bullets flew. The "beautiful Belle," with her "dewey lip and shining hair," remains "undaunted" while attending to "gaping wounds." She does not turn aside from the soldiers "waiting to die," who offer her a "voiceless prayer."[25] This image of women as angels on the battlefield persisted throughout the war in the popular press, and a visual representation of a winged nurse was even used as the centerpiece of the USSC's official emblem.[26]

Although the image of women as angels served to celebrate women's achievements, it obfuscated the gritty and gory realities of their everyday work and the work of the USSC more generally. The USSC also included a number of doctors who treated wounded soldiers and supported sanitary measures, but they slowly drifted away from the task at hand. The urgency to keep camps clean and organized enabled women and civilian doctors to enter these spaces, but as time unfolded, many doctors, operating under the guise of understanding the spread of infectious disease, began to engage in a set of systematized practices that codified racial difference.

Ever since the international slave trade had developed in the seventeenth century, some physicians and observers had put forth theories about racial difference, often postulating that Africans were naturally inferior to Europeans. Some also claimed that they were more susceptible to certain diseases, and immune to others.[27] While this line of thinking continued throughout the eighteenth and nineteenth centuries, gaining popularity at the height of the pro-slavery and abolition debates among those trying to provide a medical and scientific justification for slavery, the abolition of slavery should have ushered in a movement to dismantle these rationales. Emancipation provided an opportunity to end the underlying assumption of racial difference that justified slavery.

President Lincoln issued the Emancipation Proclamation in 1863 as a wartime strategy to drain the Confederacy of its labor force; it led to hundreds of thousands of enslaved people liberating themselves from plantation slavery and finding refuge behind Union lines. The proclamation also expanded the Northern Army's military goal from preserving the Union to emancipating slaves throughout the Confederate South. An enormous influx of formerly enslaved people arrived in Union camps. The military employed some of the men as laborers, enlisted a fraction as soldiers, and left the majority, particularly women and children, unemployed. Without proper shelter, clothing, and food, many formerly enslaved people became sick and died at the moment of freedom.[28] Rampant disease outbreaks among formerly enslaved people spread to Black troops, who were treated by members of the USSC. The Black regiments were composed not only of recently freed enslaved men but also of free Black men from the North, who joined the Union ranks and became part of segregated regiments. Together, these two groups made up an estimated 180,000 Black men serving in the Union Army.

The presence of Black troops in the Union Army led many white doctors to begin to regard race as a factor in the spread of disease. This undermined the sanitary practices that Nightingale had promoted. Although Nightingale believed in racial difference, regarding the English as the finest race on the planet, she did not use race as an explanation for the spread of cholera or other infectious diseases. Even after germ theory became widely accepted, she insisted that unsanitary environments led to disease. She did not believe that the source of disease transmission could be found in innate characteristics of the patient, even after Koch had shown that some people were carriers of cholera bacilli. Similarly, while Gavin Milroy and other doctors working in the Caribbean certainly harbored racist beliefs, they too searched for the cause of disease in the natural and built environment. Milroy condemned Black people's living conditions and blamed their high rate of illness on their failure to maintain clean homes, but he did not focus on racial difference as the cause of disease spread.

Although Milroy and Nightingale did not implicate race in the spread of infectious disease, some British doctors did take note of racial categories. In his investigation of the 1845–1846 yellow fever epidemic in Cape Verde, for example, James McWilliam identified the race of each person he interviewed, describing them as "mulatto," "dark mulatto," or "negro" (see Chapter 3). In his effort to locate the origin of the epidemic and determine its cause, however, he did not use these categories in an explanatory way. When British medical authorities and military officials read and reported on McWilliams's study, they also did not make any claims about racial differences in disease susceptibility.

Despite this trend in Britain, USSC doctors changed the direction of the early development of epidemiology as a field by resurrecting an outdated theory about racial identity as an explanation for the cause of disease. They believed that gradations of race had explanatory power. The USSC shifted attention away from the built environment as a key factor in investigating the spread of disease and focused instead on racial identity as a central point of investigation. The organization began a massive effort to catalogue the height, weight, and other characteristics of both Black and white soldiers, in an attempt to see if they differed between the races, and if these differences were correlated with different rates of illness and death.

During the war, the USSC sent agents into the field with questionnaires that included over 191 questions relating to the physical environment—the placement of the camps, the water supply, and drainage—as well as questions about food, availability of liquor, and the background of military personnel.[29] This was a similar approach to that of British bureaucrats, who used questionnaires in the field to study health conditions within the empire. The USSC collected an estimated 1,400 reports that were then analyzed in order to determine how best to improve camp life and support troops.[30] This investigation then took an odd turn. Frederick Law Olmstead, who served as the general secretary of the USSC, along with E. B. Elliot, the actuary for the commission, required doctors to start measuring and weighing white soldiers in order to compare those born in the United States with foreign-born troops. They believed that collecting this information could help the military determine which branch of military service was best suited for certain geographic locations based on the physical characteristics of its troops. During this period, some physicians believed that height served as a useful metric to gauge nutrition and health.[31]

The USSC's classification for comparing native white, foreign-born, and Black soldiers to explain varying rates of mortality and susceptibility to disease fed into a larger discourse that used racial differences as a way to justify slavery and other forms of oppression. Although Irish people were subjected to discrimination in nineteenth-century America, the USSC did not identify them as a separate group; rather, they were grouped with other immigrants into the "foreign-born" category. The evidence that the USSC collected about foreign-born troops had little or no traction after the war ended; it did not inform scientific arguments about ethnic difference, nor did those in power refer to these data to incriminate foreign-born whites later in the nineteenth century, as they did for Black people.[32]

The USSC, which had been conceived mostly as women reformers' response to the sudden, alarming outbreak of infectious disease and the helpless condition of wounded soldiers, morphed into an organization that collected data aiming to reify racial difference. The USSC sent military doctors a questionnaire, "The Physiological Status of the Negro," whose questions were based on the belief that Black soldiers were innately different from white soldiers. This assumption completely undermined the premise of equality that animated antislavery activism, which was spearheaded by women reformers. The first question was, "What facts lie at the foundation of successful efforts for preventing needless sickness?" The rest of the questions focused more heavily on the notion of racial difference. They ranged from queries about the "causes" that prevented Black soldiers from "resist[ing] disease" to others about innate immunity, hygiene, and vulnerability to specific "pathological lesions" that only affected Black soldiers. There were also questions about Black soldiers' psychological health, couched in the nineteenth-century parlance of "nostalgia" and "nervousness."[33]

The questionnaire also distinguished gradations of color among Black soldiers, asking doctors to compare how "pure Negroes" differed from people of "mixed races" and to describe "the effects of amalgamation on the vital endurance and vigor of the offspring."[34] Concerns about mixed-race people developed before the Civil War in both medical and popular discourse. In 1843, Josiah Nott, a physician and surgeon from Mobile, Alabama, had published a polemical article in a medical journal titled "The Mulatto a Hybrid—Probable Extermination of the Two Races If the Whites and Blacks Are Allowed to Intermarry." Nott claimed that white and Black people belonged to separate species and that the offspring of such unions, whom he referred to as "hybrids," were physiologically inferior to "pure" Africans and whites. He asserted that they were "intermediate in intelligence," "less capable of endurance," and "shorter lived than the whites or blacks." He added that "mulatto women" were vulnerable to diseases related to reproduction and childbearing. In fact, they were "bad breeders" and "bad nurses." If they do have children, he claimed, the children "die at an early age."[35] Nott's ideas may have influenced the writing of the USSC questionnaire, which included the question, "What influence has the admixture of white and black races upon physical endurance?"[36]

One of Nott's lines of evidence had to do with susceptibility to disease. He noted that during yellow fever outbreaks in Mobile in 1837, 1839, and 1842, "I did not see a single individual attacked with this disease, who was in the remotest degree allied to the Negro race."[37] A few years later, Nott postulated that yellow fever was transmitted not by miasmas, but by mosquitoes—a revolutionary concept at the time.[38] Yet during the Civil War, Nott's ideas about the innate weakness of mixed-race people gained more traction than his keen insights about yellow fever. His hypothesis about race coalesced with cultural narratives among both Northerners and Southerners, who were equally fascinated by the presence of "mulattoes." In the South, the varying color of people of African descent influenced the buying and selling of enslaved people, particularly women.[39] This idea played out in many novels that nineteenth-century Northerners read, including Harriet Beecher Stowe's *Uncle Tom's Cabin*, which claimed that those who were identified as "mixed" or "fair" worked inside plantation homes, while those with "darker" skin were forced to work in the fields.[40] It also made its way into abolitionists' audiences, where white allies frequently referred to famed abolitionists Frederick Douglass and Harriet Jacobs as being of mixed race, or "mulatto." The notion of mixed-race people functioned as a cultural narrative in both the North and the South, built out of certain assumptions and stereotypes. Unlike systems of racial classification in Cuba, Brazil, and other parts of the Caribbean and South America, "mixed race" did not often exist as a separate category. Because of the "one-drop rule" that prevailed in the United States, anyone who had even one distant Black ancestor was defined as "Black"—gradations of race had no legal significance.[41]

Nevertheless, the USSC questionnaire asked physicians about the physical endurance of people of mixed race. The idea that mixed-race people were inferior grew out of an antebellum paranoia about racial mixing, or miscegenation. Those who promulgated this theory attempted to dissuade the races from mixing by arguing that the Black race would be absorbed by the white race and go extinct.[42] Benjamin A. Gould, an astronomer, served as an actuary for the USSC during the war. He supervised the collection of vital statistics on soldiers and published the results in a several-hundred-page book, entitled *Investigations in the Military and Anthropological Statistics of American Soldiers*, that provided charts summarizing differences among Black, "mulatto," and white soldiers in a number of categories, including lung capacity.[43]

The USSC even became interested in measuring differences in the amount of hair that Black and white soldiers had on their bodies. Ira Russell, a doctor from Massachusetts, went so far as to observe 2,129 Black and mixed-race soldiers while they were bathing and note their level of "pilosity," from zero to ten.[44] This project underscores the gradual way that the USSC infused scientific meaning into racial identity. The mission of wartime sanitary science, as British medical authorities conceived it during the Crimean War, was to focus on the physical world to alleviate the spread of disease; but Russell's spying on Black soldiers' naked bodies during the Civil War upended that approach and used sanitary science to camouflage his salacious fascination with Black bodies.

Russell provided the most comprehensive data on Black and mixed-race soldiers. He found little evidence that mixed-race soldiers were physically inferior to "pure" Blacks; in fact, his evidence indicated that mixed-race people were taller, which suggested that they were healthier. Russell then collaborated with J. D. Harris, one of the few Black Union doctors who served in Richmond, who argued that "admixture of the races" did not "impair physical endurance" or reproduction. Harris argued that it actually promoted it. But many doctors continued to harbor the belief that there was a connection between racial identity and susceptibility or immunity to disease.[45]

The cultural fascination with mixed-race people on both sides of the Mason-Dixon Line provided the basis for a reconciliation between

Northern and Southern physicians after the Civil War. Russell had visited the Confederate doctor Joseph Jones, who pointed to his own mixed-race children as evidence that the white race would absorb the Black race, which Jones deemed as inferior. Russell wanted more evidence, so he spoke to a slave trader who had run the slave auction house in Richmond. The slave trader told Russell that mixed-race people were considered more expensive and were bought to serve as a "hotel waiters, house servants, and mechanics." He added that women who were mixed were often sold to be used as sex workers because of their physical beauty.[46]

Russell combined this evidence with his own evaluations of soldiers in order to claim that mixed-race soldiers were not inferior, as many in the antebellum period believed. Like Holroyd and McWilliam, he interviewed people, in this case slaveholders, to understand how disease spread, which expanded the role of the physicians to that of an investigator engaged in an epidemiological analysis of the factors that led disease to spread.

Unlike his British predecessors, whose interviews advanced medical knowledge and epidemiological practice by offering insights that challenged medical dogma or piloted new methods of gathering data, Russell's interviews reified the antebellum practice of using invented racial categories as valid metrics in the study of infectious disease. He drew on anecdotes from slaveholders who commodified human beings. Their ideas in turn became his ideas, and because he was a doctor working for the federal government, those ideas became the way that the medical profession and the federal government classified Black people. The Civil War ended slavery, but the use of racial classification in a deliberate twist became part of how the USSC chose to frame illness.

Unlike his British counterparts, Russell further used race as a category of medical analysis when he examined Black troops under his charge. After reading reports from two hospitals that treated Black soldiers where he worked in Benton Barracks, Missouri, Russell noted that "pneumonia is the bane of the race."[47] Russell was not the only doctor to prioritize race as a way to analyze the spread of infectious disease. Black soldiers, claimed Dr. George Andrew, an inspector for the USSC, had innately weaker lungs, which caused "diseases generally, and those of the lungs particularly," to progress more rapidly.[48] A number of physicians scattered throughout the country sent in their replies to the USSC queries, which were compiled in a "Report on the Diseases of the Colored Troops."[49] Many of their responses did not survive or were filed with other bureaucratic records and became lost in the voluminous USSC archives, which includes hundreds of boxes of documents.[50] Despite the loss of many of their responses, we know that the USSC developed questions that asked doctors to think about Black troops differently from white soldiers and established a rhetorical framework that encouraged physicians to use race as a valid biological category.

This kind of diagnosis undermined the premise of early epidemiology. Many doctors in other parts of the world were turning to the physical world and the built environment to understand how disease spread; they observed symptoms in a patient and then turned outward to housing, sewers, drainage, and crowded conditions to understand why patients were sick. USSC surgeons did the opposite. They turned inward to the patient, trying to find the answer to the illness within or on their body. While they considered the natural or built environment, they emphasized racial identity as the cause. By arguing that weak lungs caused high mortality among Black troops, rather than the inadequate housing, clothing, and other conditions that defined Black soldiers' experience during the war, physicians blamed the Black troops themselves for their pulmonary problems. This then contributed to the widely held belief that massive outbreaks of tuberculosis resulted from Black people's poor lung function, an idea that persisted for decades.[51] The Civil War ended the institution of slavery, but the USSC resurrected slaveholding ideologies to amplify racial difference and to contribute to medical knowledge.

Drawing on additional antebellum Southern ideas that grew out of slavery, USSC officials characterized Black troops' understanding of their health as primitive. Benjamin Woodward, a surgeon with the 22nd Illinois Volunteer Infantry, argued that Black soldiers' belief systems prevented them from resisting disease, and that Black people could "rarely rally from the attack of severe disease" because they believed that their time had come to die. He even reported that Black soldiers were "imbued with fatalism from the Fetish doctrines of [their] ancestors" and that spirits had the power to "trick," "poison," and "witch" them.[52] By emphasizing Black people's cultural beliefs, he further marked Black soldiers as different and primitive.[53] While many Black soldiers were certainly superstitious and held onto African beliefs and cultural narratives, many were also devout Christians.[54] USSC physicians made little to no reference to the ways in which many Black soldiers adopted Christian beliefs as a way to frame death or illness; instead they highlighted cases that portrayed Black people as primitive or different. They used religion as a way to further underscore that Black people belonged to a separate race. Claiming Black soldiers' dying words as superstitious evoked antebellum stereotypes that continued throughout the nineteenth century, most notably in Mark Twain's satirical representation of Jim in *Huckleberry Finn*.

Using their medical authority, USSC doctors assigned a medical diagnosis to Black soldiers that undermined their efforts to gain full equality. Joseph Smith, medical director for Union troops in Arkansas, wrote in 1865 that the Black soldiers "have not the intelligence of the white troops and must be made to take proper steps for the police of their persons and camp. His moral and intellectual culture is deficient and the lack of this culture renders him unequal to the white soldier in the power to resist disease."[55] By the mid-nineteenth century many doctors, from India to London to New Orleans, had moved away from the argument that morality, intelligence, or social status could explain disease susceptibility, but USSC doctors in the United States lagged behind these advances and fell back on earlier medical beliefs.[56] Smith's statement follows a logic that many in the rest of the Western world had abandoned in the wake of new epidemiological understandings about the spread of infectious disease; more significantly, it uses morality, intelligence, and culture as metrics to make an argument about disease causation. USSC doctors misused their medical authority to make claims about Black people's lack of fitness for freedom, rather than using their expertise to make diagnoses that considered the effects of living in a war environment plagued by battle, disease, and death.

The USSC's official reports about Black health revealed more about culture than medicine. But because they were framed as scientific investigations, the prejudices and cultural narratives that informed the reports became part of a medical doctrine about infectious disease. Ira Russell continued to claim that Black soldiers were different from white soldiers.[57] He argued that they suffered disproportionately from tuberculosis but were just as likely to suffer from malaria, which undermined the antebellum belief that Black people had a natural immunity to malaria.[58] The importance of his conclusions lies less in his specific findings than in his insistence on placing white and Black soldiers in different categories, which then codified notions of racial difference in relation to infectious disease.

Although Russell traced the outbreak of disease among Black troops to poor conditions, he believed in racial differences as the main explanation.[59] Russell was essentially the opposite of Nightingale in terms of the factors that he drew on to understand the spread of disease. While Nightingale made comments about racial inferiority, she emphasized unsanitary conditions as the leading cause that spread disease. In contrast, Russell commented on unsanitary conditions, but emphasized racial identity as the leading cause that spread disease. He claimed that overcrowding and exposure compromised Black people's innately poor constitutions. He sympathetically pointed out that conditions during slavery, especially poor diet, further impaired their health. And in a surprising turn, he noted that escape from slavery caused a medical crisis for enslaved people since they lacked food, shelter, and other basic necessities.[60] Although his argument has components that are similar to that of Nightingale, he nonetheless operated under the assumption that Black soldiers were innately inferior to white soldiers.

Russell further confirmed his notion of difference by performing autopsies on the bodies of dead Black soldiers. He organized his results in a table that included the height, weight, color, and age of each soldier. He then took meticulous notes on the size and weight of the heart, lungs, spleen, liver, and brain. He noted, for example, that Black troops had smaller brains than white troops, which offered scientific evidence of their inferiority.[61]

This comparison of brain size supported antebellum beliefs about Black inferiority. Samuel Morton, a physician and expert in the growing fields of craniology and ethnography, studied alleged differences in intellectual capacity among human races by measuring the size of skulls.

After evaluating over eight hundred skulls, he claimed that both Africans and Native Americans differed from Caucasians. In his book *Crania Americana,* published in 1839, Morton argued that Caucasians had the largest skulls and thereby possessed the greatest intellectual capability, which placed them at the top of the human ladder. In comparison, "In disposition the negro is joyous, flexible, and indolent" and represents the "lowest grade of humanity."[62] Morton's book advanced scientific racism and served as a justification for slavery for many in the United States, particularly in the antebellum South. Morton's ideas gained wide appeal among scientists. A few disagreed with him—not for having cast Africans and Native Americans as subordinate to Caucasians, but for failing to make distinctions between men and women.[63] A small cadre of Black intellectuals in Philadelphia, where Morton was based, refuted his claims by offering evidence of Black intelligence and equality.[64]

While these ideas gained validity among scientists and intellectuals in the first half of the nineteenth century, they were not connected with theories about the spread of infectious disease. Russell and other USSC doctors who measured and weighed Black troops' body parts took up Morton's theories and used them as a way to understand differences between Black and white mortality during the war. They applied a polemical theory to help understand a medical crisis. They upended the growing medical trend to focus on external factors as the cause of disease spread and reinvigorated an older belief that infectious diseases were caused by internal characteristics. Unlike their British counterparts, they turned to craniology as part of their so-called sanitary method, infusing a well-established racial tradition into the work of the federal government's effort to protect the health of soldiers and the broader public that extended beyond army camps.

The Civil War enabled Russell to resurrect the promise of ethnography by offering him a vast number of corpses to study. Morton scoured the globe in search of skulls, requesting amateur skull collectors to send him specimens, but the outbreak of the Civil War provided Russell with the legal, government-sanctioned opportunity to advance this practice; he performed an astounding eight hundred autopsies at Benton Barracks in St. Louis, which is roughly the same number of skulls that Morton had collected.[65] The stated mission of the USSC was a scientific investigation into sanitary conditions, but Russell turned away from concerns about cleanliness, order, and hygiene to performing autopsies as a way to gain scientific knowledge. Before the war, many doctors and their apprentices searched for corpses in order to study the human body, and many obtained them by unethical and illegal means; enterprising physicians often preyed upon enslaved people's dead bodies, which were the least protected in antebellum America.[66] Based on his detailed measurements of corpses, Russell argued that Black people were different from white people, an argument that amplified Morton's scientific racism.

The USSC had become complicit in hardening ideas about racial inferiority.[67] It further enabled Russell, a Massachusetts doctor, to engage in broader global discussions about the relationship between race, health, and medicine. Because Russell was invested in the idea of racial classification, particularly how people who were identified as "mulatto" differed from white people and "pure" Black people, he engaged the work of an unnamed "professor" who studied "mulattoes" in Brazil. The professor had claimed that "mulattoes" had superior constitutions to "pure blacks" and were "healthy, active, not dying at all." Based on his wartime service in Missouri, Kentucky, and Tennessee, Russell corroborated that theory, claiming that "mixed race" people had "exceeded the pure blacks" who had been recruited. To further support his argument he quoted a doctor from Ohio, who cited a physician in Santo Domingo, who made a similar argument about the superiority of mixed-race people in responding to infectious disease. And he also cited a physician in Cuba who had collected data that offered no evidence that "mulattos" were innately weak.[68]

The outbreak of the Civil War enabled Russell to contribute to discussions and studies unfolding throughout the Atlantic World about the spread of infectious disease in relation to racial classification. His observations not only codified ideas about racial difference in the United States but also revealed how slavery had provided the social arrangements for doctors in Brazil, Cuba, and Santo Domingo to make arguments about medicine based on subjugated populations. Throughout their reports, these physicians documented their observations and gathered statistics but were silent about how slavery created the opportunities for them to study these people in the first place. Plantations produced captive populations of people who could be studied. Slavery also enabled white slaveholders to rape, intimidate, and oppress enslaved women, which led to the creation of the mixed-race people called "mulattoes"; these people became the subjects of medical study, but the forces that produced this new population remained unstated.

Physicians' fascination with gradations of color became a way to further theorize about Black people's health, innate weakness, and physiology—it expanded their study and led to ways for them to emerge as experts. Working as a physician in Massachusetts, Russell had limited experience with global discussions about medicine or with treating Black people, but the Civil War exposed him to large populations of Black patients that turned him into an expert. In his monthly report for January 1864 of sick and wounded Black soldiers in Benton Barracks, Missouri, he described a recently organized regiment of "all shades of color from the African black to the red haired blue-eyed Anglo Saxons." His fascination with gradations of color reinforced the USSC's belief that racial categories mattered. In fact, he scribbled out what appears to be the term "Caucasian" and replaced it with "Anglo Saxons" in an effort to be more precise.[69] Before the war, Russell performed house calls and treated patients, and his intellectual engagement with the world of medicine outside of his practice likely never extended beyond writing letters to colleagues or engaging in informal discussions.[70] At the time, professional organizations were just beginning to develop; the American Medical Association had been founded in 1847, less than two decades before the start of the war.

The Civil War provided Russell and other USSC doctors with a professional status and network that enabled them to advance their theories about disease causation. Like British physicians stationed in the Caribbean and India, who postulated their theories about medicine within a military and colonial bureaucracy, the USSC created a massive bureaucracy that allowed for military physicians to develop, promote, and publicize their ideas. Russell became invested in questions about disease transmission and racial classification, and the USSC provided him with a network and audience to promote his ideas. His letters and reports functioned as part of a subregime of knowledge production that encouraged military doctors and USSC officials to accept the notion of racial difference. Similar to the cases of British physicians discussed in previous chapters, his ideas evolved from these informal contexts to formal professional forums. After the war, Russell published several articles in *Medical and Surgical History of the War of the Rebellion* and other medical journals. In the late nineteenth century Russell's research gave support to the notion of racial hierarchies that provided the roots for eugenics and scientific racism.[71]

Russell was not the only doctor whose wartime experience led him to become an expert after the war ended. Elisha Harris began his work as a doctor committed to sanitary reform in New York City in 1855. When the Civil War broke out, he emerged as a leading official in the USSC. After the war ended, his commitment to sanitary conditions continued; he was one of the founders of the American Public Health Association, serving as the first secretary and fifth president.[72] Benjamin Gould, who published *Investigations in the Military and Anthropological Statistics of American Soldiers*, was later cited by Charles Darwin and others who endorsed racial difference.[73]

✦   ✦   ✦

The USSC's efforts to document black soldiers' physical characteristics did not merely represent the work of curious doctors intrigued about racial differences. It had severe political consequences. During a war fought to liberate Black people from bondage in the name of equality, the USSC created reams of data purporting to show that Black people were innately inferior. The USSC created reports that substantiated the racialist ideology that had justified Black people's enslavement in the first place. Since the seventeenth century, doctors and others in power had regarded Africans and people of African descent as inferior, but the federal government had not previously made a medical argument about race. It had, of course, made political and legal arguments, from the three-fifths clause of the Constitution to the Dred Scott decision, that had marked Black people as inferior to white people. Yet leaders of the USSC, which was created by federal legislation, posited a medical argument about Black people as inferior. The USSC had unprecedented power to probe, evaluate, and even perform autopsies on Black people's bodies. The results of its work emerged precisely at the same time that a competing discourse about political rights and suffrage was beginning to intensify after the Civil War. Although the federal government eventually passed amendments that granted Black people birthright citizenship and suffrage, the leaders of the USSC had put forth a medical argument about Black inferiority based on science. The statistical evidence they collected aimed to provide empirical proof of inferiority. Before this time, claims about inferiority had certainly surfaced, but because they emerged as observations and anecdotes even when they appeared as laws, they could be contested.[74] The USSC, however, assigned scientific authority to medical case studies that provided evidence to stabilize the racial order after emancipation.[75]

Additionally, the USSC's form of cataloging influenced the development of vital statistics. The collection of data broken down by race furthered the use of such classifications in studying the spread of infectious disease. Nightingale had made claims about differences between South Asian and British troops, but it was not her main focus, nor was she entirely committed to making an argument about racial hierarchies. The USSC's insistence on this form of data collection is one of the reasons why it seems almost natural today, particularly among scientists, epidemiologists, and public health authorities, to create statistics that classify human beings into racial categories to understand the spread of infectious disease.[76]

USSC officials' work on racial classification and scientific racism also created a discourse that obfuscated women's work. The questionnaire set up a rhetorical framework that further amplified doctors' ideas and observations and left little room for women reformers to articulate their observations and ideas. Unlike Nightingale, who was given an official platform to articulate her theories, women in the USSC lacked a platform. After the war, many wrote memoirs and published their correspondence in order to record their work and draw attention to their contributions.[77] But during the war, the questionnaire privileged men's voices and did not offer the rhetorical space to articulate the work of women's sanitary reform. The insistence on racial difference connoted a need for scientific authority that only further emphasized doctors' knowledge.

CHAPTER 7

"Sing, Unburied, Sing"

*Slavery, the Confederacy, and the Practice
of Epidemiology*

MAYBE THE CHILD WAS CRYING. His mother held his hands as the overseer tried to take him away, pulling him from his home that resembled a hut to a clearing, far from where the cotton grew, to under a tree where two white men stood—an owner, who may not have known the child's name but could easily size him up and assign a price, and a doctor, who may have fidgeted, who maybe wasn't sure if the procedure he was about to perform would work.[1]

Or maybe there was no doubt or resistance. The Civil War had already broken out, and an unexpected enemy had emerged: disease. So maybe the child's mother did not scream because she thought this could protect them against the invisible enemies that threatened their bodies. In this scenario, she confidently—and possibly even proudly—escorted her child to the two men who stood under the shadow that the tall hickory tree cast, where cotton whirled, like snow flurries, far from where it grew, but stuck to her hair, her dress, and her child's hand. The overseer trailed behind her, because he had told her that this was going to save her child; that she was lucky to have been selected. Others had gone before them, but not everyone got the chance to meet the doctor and the owner. The doctor had no apprehension. He was fully confident because he had read an article explaining how to keep smallpox from spreading. And so the owner didn't fidget or even worry because he knew that he was not only protecting the value of this young boy but of all the enslaved people on his plantation, as well as his own family and those who lived nearby.

The doctor then reached into his medicine bag and lifted out a kit containing a tin box with scabs that had been scraped off someone who had previously been vaccinated against smallpox. The tin box also included some "lymph," the colorless fluid that oozed from under the vesicles that formed after vaccination, either in a small tube or dried between glass slides.[2] Also inside the kit was a lancet, a surgical knife that the doctor used for piercing the vesicle or scraping off the scab.

The doctor approached the healthy child and his mother with the lancet in his hand. The mother may at first have flinched at the sight of the sharp knife approaching her and her child. The child may have cried, and the mother may have begged for the doctor not to perform this procedure. But during slavery, enslaved people's crying pleas were ignored. The doctor grabbed the child. The overseer and the owner may have pinned the child against the tree and pushed the mother to the ground. The doctor grasped the child's arm and used the lancet to pierce the flesh. Using a quill, the doctor removed the lymph from the kit and placed it into the bleeding arm of the child.

+ + +

During the US Civil War, smallpox infected large numbers of people. Long before the war began, doctors and municipal officials typically quarantined patients infected with smallpox to prevent further spread of the virus.[3] While this was generally an effective preventive measure, it was nearly impossible during the war due to the vast population movements. The mobilization of troops, the emancipation of enslaved people, and the dislocation of civilians caused the virus to spread and prevented medical officials from establishing helpful protocols.[4] By the mid-nineteenth century physicians had come to rely on vaccination to prevent the spread of smallpox, but the outbreak of the war resulted in vaccine shortages as well as difficulty transporting vaccine matter across long distances. When it was not possible to obtain vaccine matter from central depositories, physicians would harvest a scab or lymph from a pustule on a recently vaccinated person and then either administer it directly to another person or store it in a vaccination kit to use later.[5]

Before the vaccination procedure was developed in England by Edward Jenner in the late eighteenth century, a different procedure, called variolation, or inoculation, was used to confer immunity to smallpox. Whereas vaccination involved treatment with milder cowpox virus, variolation involved inoculation—usually through a scratch or cut on the arm—with material from an actual smallpox pustule. Variolation had been used in Africa, China, and India for centuries, but the origins of this practice in the United States can be traced to slavery in eighteenth-century Boston.[6] An enslaved African named Onesimus introduced the famed Puritan minister Cotton Mather to inoculation as a way to prevent smallpox infection. Onesimus showed Mather the scar on his arm where he had been inoculated in Africa, and explained that it was a common practice there. In the summer of 1721, a smallpox epidemic ravaged Boston, causing fear and terror among its residents. Intrigued by Onesimus's description of inoculation, Mather, a leading official in the city, believed that this new idea came from God. He advocated for the adoption of this practice in the face of the epidemic. Many physicians balked because they believed the practice was an African superstition, and, in addition, they refused to accept a medical idea from an enslaved African. One doctor, however, was curious about the practice. Zabdiel Boylston used his own son and two of his enslaved people—a man named Jack and Jack's two-year-old son—to experiment.[7]

According to Boylston, the two children both developed fevers on the seventh day. By the ninth day, smallpox appeared on both, and they recovered a few days later. Jack, however, showed only a mild response. He did not contract a fever, and only a few pustules appeared on his arm. Boylston tried the experiment on an enslaved woman named Moll, whose response was similar to that of Jack. The doctor then tested two white men, who developed a mild form of the disease. Boylston deduced that Jack and Moll must have had previous exposure to smallpox, which is why their bodies did not respond strongly to the inoculation. He later ` inoculated dozens of others, including enslaved people, whites, and Native Americans.[8]

In the book he published on his work, Boylston provided evidence to prove the efficacy of inoculation. Although Boylston did not make any claim about racial differences among his patients, racism was not entirely absent from this episode. Slavery influenced how Boylston, Mather, and others understood the use of inoculation as a preventive measure, since Africans and people of African descent were available to doctors who wanted to observe how inoculation worked on their bodies. Located within their households, Moll, Jack, and Jack's son could be subjected to a medical experiment that their owner could pursue without fear of legal recourse or punishment.[9] Slavery also placed Onesimus in a subservient position where his only opportunity to reveal his medical knowledge was through his owner. Lastly, the international slave trade created the routes that allowed medical knowledge to be carried in the incarcerated bodies of enslaved Africans, like Onesimus, to the New World.[10]

The enslaved people in Boylston's study helped to show how human bodies react to the procedure; they made inoculation visible. Boylston then observed and documented the various stages, and, along with Mather, received the credit. Inoculation became a new medical practice in the British colonies of North America because of slavery.

In 1796, the British physician Edward Jenner developed a new, safer technique, based on the observation that dairymaids developed an immunity to smallpox after being exposed to cowpox, a similar disease that affected cattle. Jenner thus introduced vaccination (after *vacca*, Latin for "cow") using cowpox virus to replace variolation with smallpox virus, changing how many in America, Europe, and other parts of the world attempted to prevent smallpox epidemics.

The practice of administering smallpox vaccinations increased exponentially during the American Civil War because of the wartime demand for preventive measures to stop the virus from spreading among plantations full of enslaved people and army camps full of soldiers. Many Southern physicians had not performed vaccinations before the war, since there had not been any major smallpox epidemics. The wartime epidemic was the first time many of them were administering vaccines for smallpox or treating the disease.

While the precise details of some of their procedures remain somewhat unclear, in part because of a lack of standardization of terminology, the Confederacy's decision to vaccinate the entire army gave physicians an unprecedented opportunity to observe the effects of vaccination across a large population.

Not every doctor in the South knew how to correctly administer vaccinations or what to do when they triggered other medical problems or were given to people with preexisting conditions, such as exposure, gangrene, or malnutrition. Doctors in the Confederacy consequently looked to their counterparts outside of the South for help and drew on case studies published by physicians in Europe and throughout the British Empire. This further underscores how the circulation of case studies, which appeared in medical journals, treatises, and pamphlets, created a global network of physicians, who contributed to the development of epidemiology.

+ + +

Under the hickory tree, the doctor, the owner, and the overseer were injecting lymph into the crying enslaved child's arm not primarily to protect the child from smallpox, but rather in order to use the child's body to harvest lymph for more vaccinations. The war had caused an immediate need for vaccine matter. Some vendors sold it from their New York dispensaries, but Northern blockades prevented Southerners from getting reliable amounts imported.[11]

Facing this shortage, Southern doctors turned to children as the most effective way to obtain material for vaccinations. Even though the use of humans to harvest vaccine matter had proven to be dangerous— infections could result from the procedure—they nonetheless reverted to this practice, treating children as objects. Physicians had considered using soldiers, but Joseph Jones, a doctor who served in the Confederate Army, argued against this because soldiers' bodies had endured "fatiguing campaigns," "hardships," and "irregularities of habits." "Young healthy children," Jones contended, offered the healthiest bodies from which to harvest vaccine matter.[12] Children, especially those who were orphaned, poor, or from other marginalized populations, had long been used to produce material for both variolation and vaccination.[13] As I discuss later in the chapter, enslaved children were most likely used as the primary source of vaccine matter in the Civil War South. Confederate doctors may have fallen back on a proslavery parable that bondage was an idyllic institution that provided children with a healthy diet, excellent living conditions, and exposure to fresh air.

Doctors sometimes even used infants as a source of lymph.[14] Cradled in her enslaved mother's arms, an infant slept completely unaware that even her body—too young to talk or walk—had to work. The sharp, razor-like lancet pierced the newborn's paper-thin flesh and woke her, intensifying her cries.[15] The doctor came back two or three days later hoping that the infant had contracted a fever, which would signal that her body was fighting the virus and the process had started to work.

A few days later, as the mother kissed her baby's forehead to feel for her temperature or swaddled her child in a gown made from a burlap bag, she would see small red blisters beginning to erupt on her newborn's body. They could appear anywhere—on the arms, the tummy, the back, the legs, the forehead, the face, the eyelids, the back of the neck, or the bottom. Human development had set a course to determine the growth of the infant, but slavery and the Civil War had intervened; they had set the infant's body to work according to a new timeline. By the eighth day, the blisters had grown larger and fuller; they had matured into liquid-filled vesicles.[16] As the infant's body continued to labor, the vesicles continued to grow. The lymph transformed from a colorless fluid into a milky substance, turning the vesicles into pustules with a red patch around the edge. Then the doctor would slice into the vesicles to harvest the lymph. What remained on the infant's body would harden, turn into scabs, and eventually fall off, leaving a scar or a pit mark that would last a lifetime, indelibly marking a deliberate infection of war and bondage. Few, if any, knew that the scars and pit marks actually disclosed the infant's first form of enslaved labor, an assignment that did not make it into the ledger books or the plantation records.[17] Slaveholders and later generations of well-intentioned historians have followed the same logic of debt and credit, loss and gain, business and economy, and have failed to see how slavery's most vulnerable population was also employed.[18]

Using enslaved infants and children was not an aberration, but rather part of a larger pattern of using dispossessed populations as vaccine intermediaries. A French physician, for example, wanted to transport vaccine from China to Vietnam in the early nineteenth century, but carrying vaccine matter on glass or in cloth risked the deterioration of the lymph, so he turned to children as the carriers. He was not alone. A surgeon who worked under King Carlos IV of Spain put twenty-two male orphans aboard a ship headed to Mexico and vaccinated them so that their young bodies would produce vaccine matter. When they arrived in Mexico, they were used to vaccinate twenty-six more Mexican children, who were sent from Acapulco to the Philippines. And there were other cases in Italy and throughout Europe.[19]

Complications and errors often occurred after vaccinations. Not all doctors had experience performing this procedure, nor did they all follow the same steps. Dr. Jones, of the Confederate Army, recounted the experience of one physician who reported that erysipelas, a bacterial infection that causes red patches and blisters to erupt on the skin, developed among a group of "negro women and children" who had all been vaccinated with the same vaccine matter. The women and children later died. A "stout healthy white man" (possibly an overseer) who "was vaccinated with the same matter and by the same physician who vaccinated the negroes" became "extremely ill" from the "poisonous matter" and "barely escaped with the loss of a large portion of the muscles of the arm."[20]

This example was not the only case of problematic vaccinations; other instances proliferated. Southern doctors used the adjective "spurious" to refer to vaccinations that either did not "take" or caused serious illness or death in the vaccinated person. They proposed several possible reasons for these failures. Some vaccinations may have been administered incorrectly, or the vaccine matter could have been inactive. In other cases, the procedure may have proved unsuccessful because of the vaccinated person's pre-existing conditions. Doctors began to discover that gangrene or even malnutrition compromised a soldier's health, and any abrasion to the skin by a lancet could trigger an infection. Finally, and most alarming, there were cases in which, days after the inoculation, the vaccinated person began to present symptoms of another disease.

Confederate doctors began to notice that vaccine matter could transmit other infectious diseases, particularly syphilis, a bacterial infection that is normally transmitted through sexual contact. The initial symptom of syphilis is typically a chancre, which looks like a cigarette burn on the flesh where the bacteria entered. It can be easily overlooked. It lasts for a few weeks and then disappears, but the bacteria can remain dormant in the body for years before the second stage begins, producing a rash that covers the entire body.

Many nineteenth-century physicians were shocked and puzzled to witness people developing syphilis after being vaccinated for smallpox. As a doctor in Memphis explained in a 1866 letter to Jones, "The subject has occasioned me much thought, and incidentally some research, being on the watch for anything, however remote, at all applicable to the question. But I cannot say that I have formed any definite views regarding the question involved."[21]

Other Southern doctors did not doubt that vaccination could cause syphilis. One of them noted that among surgeons in the Union Army, "the opinion obtains . . . that in some instances syphilitic diseases have been definitely traced to the insertion of vaccine matter." This doctor also cited physicians in Italy who had made similar arguments in 1846, and added that at the Academy of Medicine in Paris in 1860, a committee of French doctors had advanced the argument that vaccine matter could transmit syphilis. The French doctors had affirmed "the consentaneous presence of two constitutional diseases in the same patient, and the contagiousness of secondary symptoms of Syphilis."[22] This is another reason physicians chose to harvest lymph matter from the bodies of infants and children; not only were they supposedly healthier than soldiers, but they did not have a sexual history that could have caused them to be infected with syphilis. Even Union doctors, employed by the US federal government that fought to end slavery, viewed formerly enslaved children as the ideal population to harvest lymph, as an 1863 incident in Missouri reveals.

In St. Louis, a Black regiment at Benton Barracks that included over eight hundred troops faced a smallpox epidemic in 1863. The regiment doctor began the process of vaccinating the troops with "virus obtained from the medical purveyor," but at least a hundred of them developed ulcerations as a result. Since the vaccination process had failed, smallpox quickly spread throughout the regiment. Meanwhile, a group of newly recruited soldiers, who formed the 9th Iowa Calvary, had arrived at Benton. Many of the new recruits had not been vaccinated. The surgeon in the camp had received lymph from New York, but after administering it to the soldiers, he discovered that many of the men's arms developed inflammation, erysipelas, or "glandular swellings." The smallpox outbreak intensified—some soldiers developed fever and became fatigued. Others noticed red pimples appearing on their arms, legs, and chest. Panicked, they took matters into their own hands, literally. They began to vaccinate themselves, using material from the sores on the arms of the previously vaccinated men; as a result, they developed large ulcerations. According to the regimental surgeon, several deaths occurred "as a direct result of the foul inoculation." The only men who stayed healthy were those who had good scars from previous vaccinations.[23]

Since Union physicians worried that the vaccine matter had caused the outbreak of other infectious diseases and left the troops even more vulnerable to smallpox, they needed to devise an immediate plan to thwart the spread of the outbreak. When the epidemic first erupted in late 1863, doctors faced "difficulty in procuring adequate supplies of fresh vaccine lymph from infants and healthy persons, through medical purveyors and otherwise."[24] And the vaccine matter that had been obtained had apparently been defective. "Such events created deep anxiety in the minds of the medical officers," wrote Elisha Harris, one of the leaders of the US Sanitary Commission (USSC), in his account of the episode. Medical authorities then turned to Black children: "Fresh virus was procured from New York and elsewhere, and a new and genuine stock of it was created by the aid of the healthy colored children in St. Louis."[25]

More than likely, this was not the first time that "fresh virus" from a purveyor was used to infect children in order to harvest more vaccine matter. These particular children would most likely have been among a group of refugees who had recently escaped from slavery in the South and made their way to Missouri, a border state between the Union North and the Confederate South. For enslaved people who fled from the South, or who were enslaved in Missouri, taking refuge in a border state like Missouri was precarious. On one hand, it offered protection from slaveholders and the Confederate Army, but on the other hand, slavery as an institution remained legal. Lincoln's 1863 Emancipation Proclamation only applied to the Confederate states, leaving slavery intact in the border states. By liberating enslaved people in states that had rebelled against the federal government, the Emancipation Proclamation aimed to drain the Confederacy of its labor force and cripple its economy. Lincoln kept slavery legal in the border states in order to keep those states—Missouri, Delaware, Kentucky, Maryland, and West Virginia—in the Union.

When formerly enslaved people arrived in Missouri in 1863, Union officials regarded them as "contraband," a term that left unsettled whether they were free or enslaved.[26] They had only the clothing they wore when they fled. They had no food or even tools to hunt or rods to fish. They had nowhere to sleep—not even a blanket to keep them warm or to cover the wet ground. The nursing superintendent at Benton Barracks noted, "The negroes lay in the woods and fields in cold weather while escaping from their masters."[27] Military officials in Missouri actually built Benton Barracks as a camp and hospital for them.[28] This was the same location where Ira Russell engaged in his experimental study of Black troops as part of the USSC's scientific study of health and medicine (see Chapter 6). Russell had also documented how unsuccessful vaccinations had produced "obstinate sores" in "raw recruits, in new and old regiments, in field and in quarters, in hospitals and among 'contrabands' and refugees."[29]

Like hospitals set up during the Crimean War, Civil War hospitals became sites where doctors like Russell observed large populations of people congregated in one setting. There, he and other Union doctors could easily observe the effects of a massive campaign to vaccinate a population and take note of the situations when vaccinations produced ulcerations. For soldiers, having a physician observe the effects of vaccination on their body was not an uncommon practice, but for formerly enslaved people at Benton Barracks in Missouri, where slavery was still legal, it reflected more about the imbalance of power than a humanitarian concern about the freedpeople's health and well-being. Freedpeople lived in no more than a string of tents lining the perimeter of the army camp—a shelter from the elements and a place to rest, get nourishment, and be protected from the Confederate Army. Meanwhile, as Russell documented, the hospitals at the barracks were overcrowded and poorly ventilated, often causing more health problems than they cured, as Nightingale had also found.[30] Consequently, there was little to prevent smallpox from further spreading among the Black population at the barracks, as well as throughout St. Louis.

Given these conditions, there was little Black children could do to prevent a Union doctor from using their bodies to produce vaccine lymph. In a state where slavery remained legal, a Northern doctor could easily justify using Black children's bodies to protect Union soldiers. In a camp where Ira Russell spearheaded a massive investigation, supported by the federal government, to measure, grope, and examine Black troops, the use of Black children's bodies in the service of medicine and health followed established military protocol. In a place where many Black people were forced to sleep on the ground, Black children had nowhere else to go. During a war that uprooted Black children from their families, separated them from their mothers, sisters, and aunts, and took their fathers, uncles, and older brothers to another part of the barracks to enlist in the army, these children had little protection from the doctor who approached them with a lancet in his hand.[31]

Like other episodes in the history of epidemiology during the nineteenth century, the narration of this procedure obscures the agency and perspective of the people being described. When Trotter, for example, referred to his findings about scurvy in the introduction to his medical treatise, the identification of the patients as enslaved Africans aboard a slave ship was erased. He instead referred to them as a "multitude of cases." In a similar fashion, the Union doctor obscured the status of these formerly enslaved children as "healthy colored children" who were probably refugees in St. Louis and had little choice but to offer their "aid." They likely did not volunteer.

The Civil War had led to the creation of refugee camps where newly emancipated slaves could be easily targeted as subjects for the protection of mostly white people. Using Black children's bodies to harvest lymph was not the result of the children's interest or commitment to contribute to this practice—even if the children were cajoled into believing that it would benefit their families and communities. This practice had more to do with how the social, economic, and political forces conspired to place the children in the most vulnerable position in St. Louis. This dynamic is precisely the same dynamic that unfolded at the bottom of the slave ship when Trotter learned about the causes and behavior of scurvy or when James McWilliam collected data about a mysterious epidemic that broke out in Cape Verde.

✦   ✦   ✧

Learning about the spread of infectious disease and developing preventive measures and treatments depended on imbalances of power such as those created by slavery, colonialism, and war. During the nineteenth century, the medical community understood this to be true. This kind of imbalance made possible Florence Nightingale's collection of statistics from India and from Crimean War hospitals, as well as Gavin Milroy's work in Jamaica and the army of physicians employed by the USSC. It also had a profound effect on Confederate doctors during the Civil War. They witnessed how the war produced a biological crisis that led to the spread of infectious disease and how it undermined soldiers' health. Like the USSC, the British Sanitary Commission, and other European medical authorities, the Confederacy recognized the need to investigate and document the spread of infectious disease. Confederate doctors studied accidental, or "spurious," vaccinations and reported on the consequences of these failed procedures.

In fact, their efforts to do so led Confederate doctors not only to detail their experiences during the war but also to provide accounts of their work before the war unfolded. Similar to the ways that military bureaucracy unwittingly exposed crucial details about subjugated peoples in the far reaches of the British Empire from the Caribbean to India to West Africa, Southern military authorities' requests for doctors to report on their work, document their observations, and explain their rationale revealed the common practice of using enslaved children to harvest vaccine matter. The actual reporting of this practice within the Confederate bureaucracy disclosed an antebellum practice that otherwise might have evaded documentation.[32]

The Civil War created a massive bureaucracy—like that of the British Empire—that recorded and systematized practices that later became archived and preserved. The Confederate bureaucracy made visible the practice of using enslaved infants and children to harvest vaccine matter. In February 1864, in response to continuing smallpox outbreaks, the Confederate surgeon general's office requested that hospital medical directors "promptly assign one Asst Surg in each of the larger cities of the Confederacy to the temporary & special duty of vaccinating gratis, in such cities & precincts, all healthy children, white & black, who have not been properly vaccinated."[33]

After receiving the order, Dr. James Bolton, who was based in Richmond, "traveled four weeks in the interior of Virginia, vaccinating whites and negroes." Bolton later "retraced his steps for the purpose of gathering the crusts"—the scabs that formed at the vaccination site and would be used for vaccinating others. In a letter reprinted in a report by the Confederate doctor Joseph Jones, Bolton wrote that "the result of this expedition was about eight hundred crusts, mostly from healthy negro children." Since Bolton's expedition throughout Civil War Virginia took place while slavery remained in place under Confederate rule, the enslaved children and their families would have had little opportunity to resist these procedures. Bolton reported that only one case presented "abnormal phenomena": the face and vaccinated arm of a "mulatto" child of "strumous [scrofulous] appearance" were "covered with impetigo."[34] As a Confederate doctor, Bolton not only had access to the bodies of enslaved children, but he also relied on the wartime bureaucracy to document any anomalies and to share his findings with his colleagues.

From an epidemiological perspective, slavery created an unprecedented built environment that confined hundreds of Black children to plantations where they could easily be targeted as objects to be used to produce vaccine material and then to be studied. By documenting how eight hundred vaccinations were used to produce "crusts," Bolton provided solid evidence that children could be used to produce vaccine matter. The large number also enabled him to document the singular case that presented abnormalities. The reference to the child as a "mulatto" illustrates a nineteenth-century medical practice of identifying people according to their Black heritage, which was also common among USSC physicians during the war.

Bolton's vaccine matter was not sufficient to meet the large demand in the army, but his epidemiological analysis continued to be studied after the war. Elisha Harris, who helped to spearhead the sanitary movement in the North during the Civil War and later became a pioneering statistician and public health expert, turned to Bolton's report in his published study of vaccination. He explained that the Confederate surgeon general "imported fresh vaccine lymph from Europe" and "at the same time commissioned Dr. Bolton and others to undertake the most careful repropagation of virus from healthy infants wherever they could be found upon plantations." Harris noted that Bolton's campaign was largely successful, explaining that "Dr. Bolton obtained virus which, for a time, yielded results that promised to extirpate small-pox and the foul ulcers. He ascertained that in the first thirteen hundred persons, mostly adults, who were vaccinated with that virus, only one failed to receive its normal operation and full protection."[35]

Although the Civil War had ended, and slavery had been abolished by the time of his report, Harris, a Northern doctor, drew on data that resulted from slavery. He was most concerned with failed vaccinations that produced "foul ulcers," which often resulted from incorrect procedures or comorbidities. Harris drew on the research of a Southern doctor who was, in theory, his wartime enemy; but the postwar concern to promote epidemiological understandings of effective smallpox vaccination methods united physicians from the North and South. They recognized that the war had produced a unique opportunity to study "spurious vaccinations" and drew on each other's findings.[36] This collaboration meant that Northern doctors carried Southern racism into their analysis. This is similar to the way that USSC doctor Ira Russell drew on antebellum slaveholders' understandings of racial classification in his description of Black people in the South—understandings that then became acceptable in postwar, federal government parlance despite the ending of slavery. Harris drew on evidence from slavery in order to advance medical understanding of vaccination. Since Harris was concerned with the factors that caused "spurious vaccinations," he was able to use Bolton's vaccinations of infants to visualize the production of vaccine matter that was not compromised by comorbidities. Harris, however, ignored the fact that Bolton's evidence directly resulted from slavery, euphemistically noting that Bolton and others "found" infants on plantations. Bolton did not find these infants; slavery had turned these infants into property at their birth. Confederate medical authorities forced their bodies to work for the benefit of "thirteen hundred persons, mostly adults."

The Civil War provided Southern doctors with important evidence about vaccination. One physician who was interested in the topic was Thomas Fanning Wood, who later became known as the father of public health in North Carolina. Wood began his career as a Confederate surgeon to the North Carolina Infantry during the Civil War. After the war ended he treated a number of formerly enslaved people in Wilmington, North Carolina, where the smallpox epidemic continued to rage until 1866. At the smallpox hospital in Wilmington, 761 patients were admitted between October 1865 and July 1866, and according to one account, "most of those afflicted with the malady were Negroes." Wood established a hospital for displaced freedpeople who migrated from plantations in search of freedom in Wilmington, where he treated over 1,300 cases.[37] Using pustules from his patients' bodies, he "inoculated himself many times."[38] He regarded Jenner as a hero, praising him in both personal correspondence and publications, and discussed the best way to vaccinate against smallpox.[39] In an article describing failed attempts to acquire vaccine matter by injecting cows with smallpox, Wood mentions in passing that children had been used to harvest vaccine matter during the war: "After attempting again and again to restore the efficacy of vaccine by careful cultivation among country children, no adequate supply was obtained, and then the variolation of cattle . . . was resorted to."[40]

Wood's account demonstrates how the use of children, most of whom were no doubt enslaved, began to disappear from medical history. Doctors like Wood were not historians. They were incipient epidemiologists who used their publications to advance medical knowledge but in the process dropped clues about other practices. In his discussion of vaccination, Wood referred to the wartime practice of vaccinating infants and children, which likely predated the war but was not widely documented and archived. Evidence of this wartime practice appears almost as an afterthought in his article.

The war, however, made Wood into an expert and provided him with a platform to describe, even in passing, this practice of using enslaved infants and children. In a 1944 article on the "formative years" of the North Carolina Board of Health, the author writes that "it was the war that gave Wood his chance" to practice medicine. Wood began his career as a pharmacist, studied as a private pupil of several physicians, and then entered the military and became a hospital steward. He attended lectures and took a medical exam that qualified him to become an assistant surgeon with a North Carolina regiment. After the war, he was awarded an honorary medical degree.[41] By treating formerly enslaved people, Wood became an expert on smallpox. As a Confederate doctor, having access to the wartime bureaucracy, which provided information about how other Confederate surgeons had used infants and children to harvest vaccine, crystallized his expertise.

+ + +

The history of harvesting human lymph on enslaved children's bodies was almost lost in the annals of history. Although Confederate doctors reported on this practice during the war, their records were stored in Richmond, which served as the capital of the Confederacy, and were destroyed at the end of the war. Dr. Joseph Jones, who served in the Confederate Army, consequently began a massive effort to reconstitute the South's lost medical bureaucratic records. Jones, who earned his medical degree from the University of Pennsylvania and later practiced in Georgia while holding an appointment as a professor of chemistry at the University of Georgia, understood that the war offered an opportunity to study the spread of infectious disease.[42] After he enlisted in the army as a private in the cavalry, he became interested in studying the spread of infectious disease, particularly typhoid and tetanus. He estimated that during a six-month period he treated over six hundred cases of disease in the "most unhealthful regions of the Southern Confederacy," and he wanted to spend his time after the war classifying the information.[43]

Some of Jones's records survived and offer insight about his work. On February 9, 1863, Jones, who was then at the General Hospital in Augusta, Georgia, wrote to the surgeon general of the Confederacy, Samuel P. Moore, that he had done an extensive study of tetanus and was beginning to conduct an "investigation of the typhoid fever of the camp." When this study was complete, he planned to turn his attention to intermittent, remittent, and congestive fevers. In his response Moore encouraged Jones to continue his work. Moore wrote that "the opportunities now offered of making a free and thorough investigation as to the nature, history and pathology" of certain types of fevers "should not be permitted to pass unimproved." He ordered him to "make a thorough investigation" and added that his study would not only advance science but would be of "the greatest practical importance to the army."[44]

This same practice of using evidence gained from the unexpected and unprecedented spread of infectious disease underpinned the work of British medical professionals in both the Caribbean and India. British authorities deployed Florence Nightingale to assist the army but then benefited from the scientific knowledge that she gained from her time abroad. Similarly, the British crown sent Gavin Milroy to Jamaica to protect their economic investments in the Caribbean but valued his scientific insights about cholera prevention. The surgeon general of the Confederacy recognized that the war had led to outbreaks of infectious diseases, and that the study of those diseases could advance science while helping to protect the welfare of the troops. His correspondence with Jones shows that some physicians in the mid-nineteenth century were not just caregivers; the war turned them into medical investigators, and their work contributed to the development of epidemiology. Their observations served as the basis for their understanding of disease.

This interest in medical investigation explains why Jones began a major campaign to reclaim the work of his Confederate colleagues. He knew that they, like him, had been instructed to use the war as an opportunity to collect information, process it, and study how various diseases behaved. Of most importance to Jones was the subject of smallpox vaccination. Reports of mishaps, unexpected outcomes, accidents, and failures were especially useful to him in trying to understand what could go wrong and how to improve the procedure. During the war, wrote Jones, "a number of deaths both amongst the troops and citizens were directly referable to the effects of vaccination." Several medical officers, he noted, wrote reports on the subject. Jackson Chambliss, who was in charge of Winder Hospital in Richmond, for example, documented a "large number of cases of 'spurious vaccination,' illustrated with the drawings of the various local diseases and skin affections."[45]

These records were destroyed at the end of the war, with the Union takeover of Richmond and subsequent fire.[46] According to Jones, the records related to "one of the most important subjects in its bearings upon the welfare of the human race."[47] Jones decided to write to former Confederate doctors throughout the postwar South and ask them to send him copies of their surviving reports. With the exception of a few Southern medical journals, the professionalization of physicians was still in its early stages. The American Medical Association had been founded twenty years earlier, in 1847 (and did not have a journal until 1883). In the same way that the war led to the creation of the USSC, which led doctors to collaborate more directly, the war also united Southern physicians under the banner of the Confederacy. The war had brought together physicians from across the South, and after the war, they collected, preserved, and consolidated their reports and joined together to create a professional network that helped to advance scientific ideas. The Civil War hardened the notion of a regional identity that connected physicians who would otherwise have been isolated.[48]

Jones's efforts to reconstitute medical knowledge produced by Confederate doctors eluded the federal government's efforts to wipe out any trace of the Confederacy after the Civil War. Jones was not secretive in his efforts to reach out to Confederate doctors. He openly wrote to them and even drew on evidence from Northern doctors about vaccination, and they cited his work.[49] Drawing on evidence provided by Confederate doctors throughout the Civil War South, Jones, like Wood, championed the science behind vaccination, adamantly supporting Jenner's technique and contending that the problems that occurred were caused by "ignorance, and inattention, of those who practice vaccination." "We have no more sympathy with the modern opponents of vaccination," he wrote, "than we have with those English physicians who attempted to decry the labors and to steal the honest dues of their immortal countryman, and who condescended to such low expedients as to caricature the process of vaccination." He noted that the Russian emperor, in 1811, and the king of Württemberg (modern-day southwest Germany), in 1818, had both mandated compulsory vaccination for all people in their jurisdictions.[50] Like many physicians in the nineteenth century, he drew on examples from around the globe to support his argument. Many white Southerners as well as the federal government, however, remained unconvinced. For Jones, the war offered evidence to prove the efficacy of vaccination and to uncover the factors that caused it not to work.

Like slavery, colonialism, and the Crimean War, the Civil War created a captive population of people confined to a specific region who could be studied. Confederate authorities led an effort to vaccinate the entire army. They instituted vaccination primarily to protect the troops against a biological enemy, but the widespread vaccination program also enabled Confederate doctors to observe how patients' bodies reacted to these procedures. They were able to test the hypothesis that poor nutrition or ill-ventilated rooms led to unhealthy bodies that responded poorly to vaccination. They were able to investigate theories about how vaccination could transmit other diseases. They were able to observe which bodies produced effective vaccine matter and which bodies did not. The Civil War created a network of physicians in the South who could share information about vaccination that had very clear textbook explanations but produced unfamiliar effects on the ground.

Confederate doctors not only studied the effects of vaccination among troops but also turned to another captive population: prisoners of war. Crowded in filthy tents, starving, thirsty, emaciated, sick, and suffering, Union prisoners provided Confederate doctors with an opportunity to observe serious problems with vaccination, many associated with scurvy. Both Confederate soldiers and prisoners suffered from scurvy. Many soldiers "manifested slight scorbutic symptoms," noted Jones, but little attention was paid to this condition when performing vaccinations because of the mildness of the disease. Symptoms of scurvy were much more prevalent in prisoner-of-war camps, where captive Northern soldiers were crowded together in a confined environment. This allowed Confederate doctors to more closely examine the situation, especially at Andersonville, a Confederate camp in southwest Georgia—which has a reputation for being the most inhumane and ghastly of all the prisoner-of-war camps. Confederate authorities had decided to vaccinate incarcerated Union troops to fend off a smallpox epidemic. After being vaccinated, however, the prisoners began developing unusual reactions. The vesicles produced in vaccinated prisoners who had scurvy, explained Jones, were large and unhealthy, but the "decomposing pus and blood" from these vesicles was used to vaccinate others anyway. "Gangrenous ulcers" then formed where the lymph had been inserted. These ulcers caused severe tissue destruction, "necessitating amputation in more than one instance."[51]

After the war ended and the prisoners were released, the United States charged the Confederacy with purposely poisoning Union prisoners of war at Andersonville by vaccinating them; one hundred men had suffered amputation, according to the charge, and two hundred died. Although vaccination was a known scientific principle, many people, both doctors and nonmedical professionals, did not understand the consequences of incorrectly administering a vaccination or the unintended effects of vaccinating a person suffering from another disease. The federal government interpreted the amputations and deaths after vaccination as evidence of a concerted effort to poison Union prisoners of war.

In addition to those who had been vaccinated, the Union army also discovered other surviving prisoners of war at Andersonville who were naked and emaciated, sick, unconscious, or missing body parts. Some had been tortured, and many, if not all, had been forced to endure without sufficient food, clothing, and shelter. Of the 45,000 prisoners who entered the prison, approximately 13,000 died, largely of disease. The United States government charged Henry Wirz, who served as commandant of the military prison at Andersonville, of conspiracy "to injure the health and destroy the lives" of US soldiers and "murder, in violation of the laws and customs of war."[52] A two-month trial, beginning in August 1865, was held before a special military commission. Wirz was found guilty and was hanged in November 1865. Wirz's trial has been remembered as one of the first prosecutions of war crimes in history. Although it was not the only one held after the Civil War, it served as a precedent for the Nuremberg Trials held after World War II.[53]

While the harrowing history of Wirz and Andersonville has been well documented, the accusation that Wirz ordered Confederate doctors to purposely poison Union prisoners of war by vaccinating them against smallpox has been largely overlooked.[54] The charges included a long list of additional abuses, including appalling punishments that range from whipping and starving prisoners to forcing prisoners "to lie, sit, and stand for many hours without the power of changing position" to setting bloodhounds loose to find escaped prisoners and then encouraging "said beasts to seize, tear, mangle, and maim the bodies and limbs of said fugitive prisoners of war."[55] Within this context, accounts of purposely poisoning prisoners with smallpox vaccine appear as yet another example of how Confederate authorities inhumanely treated Union soldiers.

The military commission also charged two doctors, Joseph White, who was the surgeon at the post, and R. R. Stevenson, who was in charge of the military hospital, as co-conspirators, along with three other men who were involved in operating the prison. Surviving POW Frank Maddox, of the US Colored Troops, was imprisoned at Andersonville, where he worked as a gravedigger. Maddox testified that the doctors were ordered to vaccinate the prisoners, and after they died, the doctors "cut some bodies open, and . . . sawed some heads open"; he remembered seeing a green streak that extended from the arm to the body in some cases. Maddox remembered hearing Wirz tell one of the doctors to vaccinate all the men. "I saw Captain Wirz in the graveyard with the surgeons two or three times," he testified; "they were laughing over the effects of vaccination one day . . . they were laughing about its killing the men so."[56]

The commission interpreted the shocking numbers of amputations and deaths among the federal prisoners as evidence of mistreatment.

Wirz's defense then called John C. Bates, a Confederate doctor, as a witness. He had been stationed at Andersonville from September 22, 1864, to March 26, 1865. He attempted to shed light on the science underpinning vaccination and offered an explanation for the high rate of mortality among the prisoners. He explained to the commission that the smallpox vaccine, which had been given before he arrived at Andersonville, is "a poison at best," but that does not mean that it should never be administered. He described the overcrowding in the hospital, the filthy conditions and vermin, which compromised the prisoners' health. He described the men as extremely emaciated, "worn down" and "cadaverous," and explained that because they suffered from scurvy, the slightest prick of the skin, such as from a lancet, could lead to gangrene. When cross-examined and asked if he would have performed vaccinations given the prisoners' ailing health, Bates explained that while the vaccine was a poison, smallpox was an even worse poison: "If those men took small-pox in their reduced condition, it was certain death. If by vaccination I could have saved one man in ten, or one man in 50, or one man in 100, it would have been my professional duty to have vaccinated."[57]

Joseph Jones, who had begun the massive effort to collect information from former Confederate doctors after the war, argued that it was outrageous to suggest that Confederate surgeons would create a conspiracy to poison prisoners and explained that the high mortality resulted from complications. As he explained, "the charge of deliberately poisoning the Federal Prisoners with vaccine matter, is a sweeping one, and whether intended so or not, affects every medical officer stationed at that post; and it appears to have been designed to go farther and to affect the reputation of every one who held a commission in the Medical Department of the Confederate Army."[58] Jones had witnessed the mortality and the unusual effects of vaccination during the war. When the Union prisoners of war started to develop dangerous reactions to the vaccinations, Confederate authorities ordered Jones to launch an investigation into that phenomenon as well as the other diseases that were ravaging the prison. He compiled an extensive study but was unable to share the full report at Wirz's trial. It had never been delivered "on account of the destruction of all railroad communication with Richmond, Virginia," and then, after the war, it was seized by "agents of the United States government." Jones sent an appeal to Norton Chipman, the judge advocate overseeing the trial, explaining that the Confederate government did not deliberately "injure the health and destroy the lives of these Federal Prisoners." The original report, he explained, was produced for the surgeon general of the Confederacy "and was designed to promote the cause of humanity, and to advance the interests of the Medical profession."[59]

As a witness at Wirz's trial, Jones turned over the surviving evidence that he had collected.[60] He asserted that he personally saw only a few cases of injury from vaccination, but in one man he examined who had had his arm amputated, "I was led to believe that it was in consequence of the condition of the system of the man, rather than from the matter introduced, from the fact that small injuries were frequently attended with gangrene in that foul atmosphere."[61] In his appeal to Chipman, he asserted that the Confederacy had developed specific policies for Union prisoners of war that protected their health and well-being. He described a Confederate policy from May 21, 1861, a month after the war began, that mandated that prisoners of war receive the same rations "in quantity and quality" as the enlisted men in the Confederate Army. Although the quantity of food might have been adequate, a diet based only on "unbolted corn meal and bacon" would cause scurvy, which, if left untreated, could cause "secondary hemorrhage and hospital gangrene." Scurvy, dysentery, and diarrhea would have been much more common among the Confederate soldiers, he contended, if extra supplies had not been made available by citizens and charitable groups.[62]

As the war dragged on, depleted supplies and lack of access to rations led to great suffering among soldiers. The suffering of the prisoners of war, according to Jones, was not unique to them but reflected larger inadequacies within the Confederacy. Jones stated that Confederate authorities wanted to accelerate the prisoner of war exchange with the Union Army in order to relieve the South of the expense of feeding and clothing the prisoners. He argued that the Confederacy's inability to manage the federal prisons could be attributed to "the distressed condition of the Southern States," which were suffering from a financial crisis, a diminished army, ruined railway lines, and "starving women and children and old men fleeing from the desolating march of contending armies."[63]

The widespread practice of vaccinating soldiers enabled Jones and his colleagues throughout the Confederacy's medical department to study failed vaccinations, which provided Jones with evidence that the prisoners had not been poisoned. Before the war, there were no major smallpox epidemics in the antebellum South—there were a few in the West, affecting mostly Native Americans, but none that would have given Southern doctors the opportunity to observe the effects of a full-blown epidemic. Since no major epidemic threatened the lives of Southerners, there were no massive campaigns to vaccinate large portions of the population. There were several smallpox outbreaks in the United States in the late eighteenth century, and doctors and municipal authorities debated the legality, medical efficacy, and religious meaning of mandating vaccinations. Some argued against this practice because it violated citizens' rights, while others advocated for it.[64] The legal questions that troubled eighteenth-century Americans did not apply to the Confederate Army during the war. Like the Union Army, the Confederacy used its authority to require vaccination.

The Confederacy's efforts to vaccinate soldiers, prisoners of war, and, in some instances, citizens and enslaved people, was likely the largest vaccination campaign in the United States in the nineteenth century.[65] This massive campaign made visible the accidents and unintended consequences of vaccination; one doctor could witness the consequences of this procedure among a large number of patients. Since many of the medical records of the Confederacy were destroyed at the end of the war, it is not clear how many vaccinations took place, but the campaign extended across the entire Confederate South, from Virginia to Louisiana. Physicians supplied Jones with their best recollections and summaries of their surviving reports to create a comprehensive statement on vaccination practices during the war, which he then published as a study.

Based on this comprehensive information, Jones confidently argued that the Union prisoners were not poisoned but rather that the violent reactions to vaccination were caused by the prevalence of scurvy and the crowded conditions of the hospital. He explained that the prisoners developed gangrene from any "abrasions" of their skin, from insect bites to splinters, because of the "scorbutic condition of their blood." The lancets used to prick their skin to perform the vaccination injured them in the same way. The scabs that then formed on these patients produced injurious effects when used to vaccinate other prisoners, unlike scabs from healthy people.[66] While it might appear odd that doctors would use scabs from patients who had an atypical reaction to vaccination, there was no understanding at the time of how germs transmitted disease; in fact, part of Jones's study explored whether lymph and scabs used for vaccination could transmit syphilis or other diseases.

To buttress his argument that it was the conditions in the camp, not poison, that had led to the spread of disease among the prisoners, Jones discussed what doctors in other parts of the world had found about the dangers of a poor diet and crowded conditions. Jones referred to Trotter's study of scurvy on crowded ships as evidence, as well as to the work of Gilbert Blane, a British naval physician who served as physician to the fleet in the late eighteenth century and advocated improvements in hygiene, ventilation, and provision of citrus juice to prevent scurvy.[67] Jones wrote, "It has been well established by the observations of Blane, Trotter, and others that the scorbutic condition of the system, especially in crowded camps, ships, hospitals and beleaguered cities, is most favorable to the origin and spread of foul ulcers and hospital gangrene."[68] In referring to camps, ships, hospitals, and cities, Jones is using shorthand to refer to research that grew out of the international slave trade, late-eighteenth-century British and European prison reform, the redesign of the Paris hospitals, and Florence Nightingale's work during the Crimean War. This list of the most commonly crowded spaces in the nineteenth century served as a standard example that doctors like Jones throughout the world used to illustrate how crowded spaces led to the spread of infectious disease.

In order to further buttress his argument that the prisoners were not intentionally poisoned but rather were suffering from scurvy, Jones cited research by a range of other doctors across the globe. He began with seventeenth-century British physicians' discussions of gangrenous ulcers that result from scurvy and continued with the observations of the British surgeon John Huxham, who studied fevers and wrote about scurvy among sailors. Jones turned to the Caribbean and discussed the work of John Hunter, a surgeon in the British Army, who wrote about sores and ulcers appearing among soldiers in Jamaica. He then elaborated on James Lind's, Trotter's, and Blane's findings and quoted a treatise written by a doctor about the Crimean War, who documented that wounded French soldiers who were transported to a hospital in Constantinople suffered from gangrene due to the overcrowding of the ships and hospitals.[69]

After putting forward this evidence, Jones concluded that the "foul scorbutic ulcers," "hospital gangrene," and complications of vaccination that occurred at Andersonville "were by no means new, in the history of medicine," having happened in various wars and sieges across time and space.[70] Both Trotter and Blane benefited from being on ships, a controlled environment that enabled them to better observe how scurvy developed. Enslaved Africans and seamen were forced into confined environments and given rations that did not include vegetables or fruits. In both cases, the populations developed scurvy. Jones tried to convey to the commission that Andersonville was analogous to a ship. "In truth these men at Andersonville, were in the condition of a crew at sea," Jones wrote, "confined upon a foul ship, upon salt meat and unvarying food, and without fresh vegetables. Not only so, but these unfortunate prisoners were like men forcibly confined and crowded upon a ship, tossed about on a stormy ocean, without a rudder, without a compass, without a guiding star, and without any apparent boundary or end to their voyage."[71] Jones argued that Union prisoners were not poisoned; when they were vaccinated they developed ulcerations because they were also suffering from gangrene or scurvy, which then led to amputation or death.

Finally, Jones provided evidence from Northern prisons showing that more Confederate prisoners of war died in Union camps than Northern troops at Andersonville. Jones cited an essay by a Northern medical inspector who contended that overcrowding in the camp at Murfreesboro combined with scurvy led to a high mortality among Confederate prisoners of war. Jones believed that there were just as many deaths resulting from vaccination among the Confederate soldiers in Northern prisons as among their counterparts at Andersonville.[72]

Despite Jones's compelling efforts to explain the science that under-pinned failed vaccinations, Judge Advocate Chipman, presiding over Wirz's military trial, disagreed adamantly with his claims, rejecting the fundamental premise of Jones's report to the surgeon general about An-dersonville. Chipman quoted from the letter of instruction that the sur-geon general sent in August 1864 to I. H. White, the physician in charge of the prison hospital at Andersonville, which ordered the staff at the hospital to assist Jones in the autopsies he planned to perform, in order "that this great field for pathological investigation may be explored for the benefit of the medical department of the confederate armies." Chipman used this instruction to argue that "the Andersonville prison, so far as the surgeon general is concerned, was a mere dissecting room, a clinic institute to be made tributary to the medical department of the confederate armies."[73] He further stated that the Confederate doctors knew that the vaccine matter was poisonous and administered the vacci-nations nonetheless. In support, he summarized the testimonies of nine witnesses, including that of Frank Maddox, who had testified that he heard the Confederate doctors laughing about having killed the men by vaccinating them. Chipman further criticized Jones's efforts to offer a medical history of failed vaccinations by asserting that "vaccination with genuine virus has never before resulted in such frightful mortality. The records of medicine and pathology nowhere, in no country and no age, afford or approach a parallel to Andersonville." He then undermined the Confederacy's argument that vaccination was a necessary preventive step, noting the relatively few cases of smallpox in the camp. Chipman contended that the doctors knew that vaccination would kill the pris-oners because they were in such poor condition, and concluded that in compelling prisoners to be vaccinated, the doctors showed "heartless-ness," "implacable cruelty," and "evil intent."[74]

The US government ultimately found Wirz guilty on two charges, including "maliciously, wilfully, and traitorously" conspiring with sev-eral Confederate doctors "to injure the health and destroy the lives" of prisoners of war being held in Confederate prisons, so as to weaken the US Army. The list of specifications under this charge included the deadly smallpox vaccinations: "And the said Wirz, still pursuing his wicked purpose . . . did use and cause to be used for the pretended purposes of vaccination, impure and poisonous vaccine matter, which . . . was then and there, by the direction and order of said Wirz, maliciously, cruelly, and wickedly deposited in the arms of many of said prisoners, by reason of which large numbers of them, to wit, one hundred, lost the use of their arms, and many of them, to wit, about the number of two hundred, were so injured that they soon thereafter died."[75]

+   +   +

Recall that up through the nineteenth century, medical authorities, from Trotter in the underbelly of the slave vessel to Nightingale in the crowded hospitals in Scutari, were developing medical theories based on their observations. Then they shared the information they had gathered. For Judge Advocate Chipman to argue that the medical profession knew that the prisoners of war suffering from scurvy might die if they were vaccinated misunderstands how medical knowledge operated during this period in the South and even throughout large parts of the world. Doctors learned more about the spread of infectious disease from their observations and experience than from textbooks or medical principles. Medical understandings were rapidly changing during this period. Doctors were examining the causes, often in the natural and built environments, that led disease to spread. Chipman was working under the reasonable but incorrect understanding that medical knowledge existed in a textbook and that these doctors ignored those principles. Instead, the broader global context—from Trotter to Nightingale— indicates that doctors learned about infectious disease on the ground, and in that way, the Confederate doctors resembled their counterparts in other parts of the world.

Jones drew on the vernacular that medical professionals used when they explained how disease spread: concrete examples, not abstract theories. Recall that Nightingale refused to accept Koch's germ theory because it was far removed from the physical world that she believed caused disease, even after she observed bacteria under a microscope. Medical knowledge grew out of what doctors observed on the ground and then reported back to officials who collected the facts, analyzed them, and then developed theories.

Chipman's condemnation focused exclusively on the Union prisoners of war at Andersonville, but there was a more pernicious practice that even his fastidious investigation into the Confederacy's medical department missed. Throughout his report, Jones contended that vaccinations were most successful when they were administered using vaccine matter taken from infants and children, most of whom were enslaved or did not appear with the racial modifier describing them as Black. While Jones referred to Jenner's work and his use of presumably white children in England, where he conducted his studies, as well as the use of children in Italy, the term "infants" in the Confederate South likely implies Black infants even when no race was indicated.[76] Recall that, as we saw in Trotter's study of scurvy in enslaved Africans, references to subjugated people of color often appear in doctors' first drafts, but once their theory becomes codified, the racial marker slips off the page. While it is not impossible that poor white infants may have been used, there is evidence that Black infants and children were used.

Judge Advocate Chipman spoke on behalf of the dead prisoners at Andersonville, but no one spoke on behalf of the infants and Black children whose bodies were used to cultivate smallpox vaccine. There was no trial for them. There were no witnesses or rebuttals or charges leveled against those who engaged in this practice. There was not even an investigation that explored the effects of these procedures on their bodies. Did they suffer from gangrene? Did a green streak run from their arms to their chests? Given the poor and inadequate diet enslaved people were forced to survive on, would these infants and children have suffered from scurvy, too?

While the answers to these questions remain unknown, the response to the outbreak of smallpox during the Civil War followed similar global patterns that advanced knowledge about infectious disease. The Confederacy, like the British Empire, created a massive bureaucracy that compelled physicians to observe, document, and exchange their ideas about infectious disease. As in other episodes throughout the nineteenth century, Confederate doctors depended upon populations of subjugated people—bondspeople and prisoners of war—to develop their theories.

The book Joseph Jones published in 1867, *Researches upon "Spurious Vaccination"; or The Abnormal Phenomena Accompanying and*

Following Vaccination in the Confederate Army, during the Recent American Civil War, 1861–1865, recounted his testimony during Wirz's trial, provided a synopsis of theories about smallpox from doctors in Europe and America, and reproduced letters and reports from Confederate doctors who wrote about vaccination complications.[77] Like other doctors across the globe during the nineteenth century, his observations on the ground served as the basis for his research. His study offered evidence of the main factors—scurvy, malnutrition, and syphilis—that prevented vaccinations from working effectively and that often led to other medical problems and even death. He also pointed out that failure to properly administer vaccinations could cause them to be ineffective. Since most Confederate doctors had never witnessed smallpox epidemics in their hometown practices before the war, the publication served as a manual for the medical profession in the event of another epidemic emerging. In this way, Jones's treatise is similar to Gavin Milroy's study of cholera in Jamaica. Both men were employed by the government, and both were commissioned to use their observations and collect testimonies from their colleagues in order to understand the spread of infectious disease.

The Confederacy's bureaucracy, like that of the British Empire, also illuminates a key feature within the study of epidemiology: the ability to see the effects of disease unfolding in various settings at the same time. In the same way that military commanders mapped strategic campaigns across a broad geography by means of reports from officials on the ground, the Confederacy's medical authorities had the ability to observe failed vaccinations across a broad terrain. This feature came into sharp focus after the Civil War ended, when a global cholera pandemic broke out in 1865.

CHAPTER 8

Narrative Maps

Black Troops, Muslim Pilgrims, and the Cholera Pandemic of 1865–1866

B Y 1885, the Civil War had been over for almost twenty years. The once explosive epidemics that had transformed Confederate military camps into makeshift vaccination centers had been forgotten and replaced with a homespun memory of noble heroes who died on the battlefield. As Confederate wives and widows planned to erect Confederate statues in Southern cemeteries to honor the dead, Northern veterans and their family members filled out tedious pension applications. They documented wartime injuries, disease, or fatalities to access federal compensation for the suffering and loss that had occurred, mostly in Union camps, where the United States Sanitary Commission (USSC) had largely failed at preventing the spread of infectious disease. Many Black veterans struggled to access their pensions.[1]

While much of the morbidity and mortality from the war had faded from public view by 1885, the knowledge that doctors had gained from the Civil War and Reconstruction era became useful to the medical community when a new cholera pandemic threatened to reach the United States. Three previous cholera pandemics had swept into the country—in 1832, 1849, and 1866. The third, which unfolded in the immediate aftermath of the war, provided military physicians with ample opportunity to investigate the cause and spread of the disease, as well as to explore preventive measures.

As the nation braced itself for another pandemic in 1885, Edmund C. Wendt, a New York–based pathologist, turned to the past to compile a treatise on cholera. He collected case studies from previous epidemics in Asia, Europe, India, and the United States from 1817 to 1883, and included chapters by himself and other physicians on the history, etiology, symptomology, pathology, diagnosis, treatment, and prevention of the disease. Section 2 of the first part of his treatise, "A History of Epidemic Cholera, as It Affected the Army of the United States," was penned by Ely McClellan, a major and surgeon in the US Army. In the introduction to the volume, Wendt praised McClellan's contribution: "Dr. McClellan's history of the epidemic as it has affected the United States Army constitutes an authentic record of a highly instructive subject. Certainly nothing could furnish a more convincing proof of the agency of human intercourse in the dissemination of cholera than this part of the volume."[2] McClellan reviewed the history of cholera outbreaks in the army and theorized about what caused cholera to develop and spread through the troops. The military bureaucracy allowed McClellan to gather together reports on cholera in the army during all of the nineteenth-century epidemics, starting in 1832.

In compiling his treatise, Wendt hoped "to furnish the physician with a faithful account of the actual state of our knowledge." Like many of his generation who investigated epidemics, Wendt relied on information from physicians around the world, including "American, English, French, German, Italian and Spanish writers who are recognized in their own countries as the highest authorities upon the subject."[3]

Wendt published his treatise about a year after Robert Koch had identified the cholera bacillus in India. Koch claimed that this bacillus was the causative agent of the disease and could be transmitted through contaminated water. Despite Koch's work, some nineteenth-century medical authorities continued to debate the cause of cholera. Wendt explains in his introduction that he has devoted a "special chapter" to Koch's discovery. He notes that "while Koch's doctrine has not been finally established as a scientific truth, it has very much in its favor. At the same time space has been given to opinions directly traversing the points claimed by the German investigator."[4]

Like many of the other doctors we have encountered in this book, Ely McClellan owed much of his knowledge of disease to war. He discussed the environment in which cholera outbreaks took place, documented illnesses and deaths, and theorized about the cause, spread, and prevention of cholera. In a previous article on a reappearance of cholera in 1873 he had put forward the argument that the disease was contagious and could be traced to the movement of individuals, but he faced a "storm of adverse criticism." In the introduction to his chapters in Wendt's treatise, however, he claims that his theory has been "strikingly verified by the experimental research of Koch and other observers."[5]

McClellan's contribution lies less in his conclusion than in his method, which enabled him to visualize the spread of cholera and to consider what social and environmental conditions could prevent it from spreading. McClellan begins his analysis with a detailed account of the cholera epidemic of 1832–1835. He documents the movement of ships, the conditions of garrisons, the number of infected individuals, the dates when cholera appeared, the time between outbreaks, and the precise locations of the cases.[6] The subjects of study were mostly troops involved in the Black Hawk War of 1832. In 1830, US imperial policies pushed Sac and Fox Indians from their homeland in northern Illinois further west to what is now Iowa, but the Indigenous people struggled to eke out a livelihood there. Their leader, Black Hawk, then led them back to northern Illinois, which triggered a war with the US Army.[7]

Military reports and imperial policy created a medical theater that enabled McClellan to conclude that the disease spread as a result of troop movements. War and dispossession had set the stage for McClellan to visualize the spread of cholera, and to theorize how it moved from one location to the next. McClellan relied on observations of Native Americans as well as army troops to observe the spread of cholera through the Middle West. He reported that it had reached Native American prisoners of war at Rock Island, who became ill after they were released and returned to their homes. The disease "continued epidemic" among them, as evidenced by the fact that troops who worked as

summer scouts "in the Indian country along and between the Mississippi and Missouri rivers, almost invariably suffered from the disease
during the years 1833 and 1834."[8] President Andrew Jackson's imperialist policies, which led to both the deployment of troops and the violent
dislocation of Native Americans, established the geographic coordinates for McClellan to track the presence of cholera in this region. The
troops, and especially the Native Americans, likely did not know that
their infection would later help McClellan map the pandemic.

During the mid-nineteenth century, before pictorial maps had become synonymous with epidemiological investigations, medical authorities like McClellan relied on written narratives to track the spread
of diseases. When McClellan could not see cholera move, he theorized
that the germs had remained infectious during a "holding over" period
and had then become reactivated. He illustrated this idea by drawing
on case studies, including one account that examined Black troops at
Jefferson Barracks, Missouri, and another that discussed the "Indians
and negroes" at Fort Gibson, Arkansas.

According to McClellan's account, in January 1867 the 38th US Colored Infantry was organized at Jefferson Barracks, Missouri; it included over 1,200 recruits, "mostly discharged soldiers from volunteer
regiments." A severe epidemic had occurred at that location in the
summer of 1866, but no cases occurred among the 38th Infantry while
they were stationed there. The troops were sent for duty to New Mexico
but were then redeployed to protect the workers along the Kansas Pa-
cific Railway from Indian attacks. It was the movement of this regiment
from Missouri, claimed McClellan, that "occasioned a most disastrous
outbreak of cholera on the high, dry plains of western Kansas."[9]

On June 28, one soldier from the camp of the 38th Infantry's Company H, near Fort Harker, Kansas, was stricken with cholera and taken
to the hospital at the fort, where he died. A civilian near the fort also
died. On June 29 and 30, two more cases were reported from Company
H and were also transported to Fort Harker. Over the next few weeks
the medical officer reported sixteen more cases of cholera at the Company H camp, five of which proved to be fatal. The first case in Fort
Harker itself occurred on June 29. On July 10, a portion of the regiment
began a march from Fort Harker to Fort Union, New Mexico, with sev-

eral new cases breaking out along the way; by the time they reached New Mexico, forty-six cases had been recorded among the enlisted men, with seventeen deaths.[10]

Initially McClellan could not determine how it was possible that the 38th Infantry became infected with cholera at the newly built Fort Harker in Kansas when there had been no reported cases among them in Jefferson Barracks since the previous year. After carefully investigating various factors, McClellan concluded that the clothing given to the troops at Jackson Barracks had probably been infected with the "cholera germ," and that the germ had spread rapidly after they reached Fort Harker due to the unsanitary conditions of the camp. McClellan provided evidence from another military physician, George Sternberg, who reported that "the police of the camps [at Fort Harker] was not good when cholera first made its appearance. Some of the company sinks [latrines] were in wretched condition, and there were several offensive holes about the post where slops and garbage from the kitchen had been thrown."[11]

McClellan witnessed how cholera spread across the central United States. Knowing about the cholera outbreak from the previous year, he could trace the origin of the epidemic to the last known place that it had surfaced, that is, Jefferson Barracks. Knowing that there were no recent reported cases of cholera in that location, he then focused his attention on the physical factors that confronted the regiment—a new location and new clothing. Because of the military bureaucracy, he had records that provided him with two new variables that had not been present the previous year. He then theorized that these factors caused the epidemic. He surmised that clothing infected with the cholera germ had led to a full-blown epidemic in an unsanitary camp. Without access to all of these variables, he or any other doctor in the nineteenth century would have had a difficult time tracing the spread of cholera. Yet, as an army surgeon with access to data provided by the military bureaucracy, he could observe the various moving parts of the event in an effort to gain a fuller understanding of how cholera spread.

McClellan's conclusion may have been wrong—we now know that cholera is not normally transmitted via infected clothing—but his contribution lies less in his understanding about medical science than in the methods that he was developing to understand how cholera spread. The outbreak among the Black troops provided him with a case study as well as key information about their health and the environments where they lived. By providing a detailed narrative account of the camp conditions and the movement of the troops, McClellan established an investigative framework that allowed him to track the spread of cholera by closely examining the environment, tabulating the number of infections, paying particular attention to the movement of people, precisely documenting dates, and drawing on evidence from other medical professionals.

McClellan was particularly interested in understanding what he perceived as the cycles of cholera: how and when it emerged, the number of people it infected, how it seemed to subside and then reappear. In the same report, McClellan also details the outbreak of cholera at Fort Gibson, Indian Territory (now Oklahoma), which, he notes, reappeared in June 1867 "among the negroes and Indians" and then spread to Company D, 10th US Cavalry.[12] McClellan carefully charts the movement of the disease from Fort Gibson to Fort Smith and then to Fort Arbuckle in Oklahoma. He records how many troops became infected and notes that two companies of the 6th US Infantry suffered a severe epidemic after they passed "over the road" and "perhaps used the camp-grounds" that had been used by Company D. This example, like many others in his report, all focus on how cholera could spread without warning.

The military bureaucracy provided him with knowledge of the social geography of the landscape where the troops resided. McClellan noticed that adjacent to the military camp a group of Black people and Native Americans lived. This could have been one of the refugee camps that were common during the Civil War period, that served as shelter for newly emancipated enslaved people seeking protection from the Union Army after fleeing from slavery. Or, based on the presence of Native Americans congregated with Black people, it could have been a type of maroon community, which developed before the war, where runaway enslaved people found refuge among Native Americans.[13] In either case, these populations enabled him to trace the spread of cholera.

✦   ✦   ✦

The roughly thirty-year period between 1854 and 1885, though marked by both clarity and confusion about disease transmission, helped to crystallize the development of the field of epidemiology. Both supporters and skeptics of the waterborne theory of cholera created ways of trying to visualize the spread of cholera. Their efforts to see an invisible agent led to the advancement of qualitative observations that tracked the epidemic around the world. Medical authorities stationed in Constantinople, Paris, London, and Washington, DC, collected narratives about cholera's movement in an effort to better monitor epidemics. Due to the rapidity with which cholera spread, they became increasingly committed to reporting on cholera in other countries in order to anticipate its movement. Before the popular use of maps, which would serve as a key technology in modern epidemiology, physicians relied on bureaucratic narrative reports to track cholera from one region to the next.

Cholera had existed in parts of Asia for centuries, but the global transformations of the nineteenth century—increased trade and travel combined with human migration—led to the disease spreading from Asia to Europe to the Americas and other parts of the world.[14] Many British and European medical authorities traced the spread of cholera to the movement of Muslims who traveled to Mecca as part of their annual pilgrimage, known as the hajj. British medical authorities claimed that when Muslim pilgrims left the Middle East and boarded ships to travel back to their homes, they spread cholera to Egypt and the Mediterranean; from there, it continued on to Europe, England, and across the Atlantic Ocean to the United States and the Caribbean. The third major cholera epidemic to affect the United States began in 1866, a year after the end of the Civil War, which only exacerbated the situation. The dislocation of white Southerners, the return of armies, and the migrations of freedpeople intensified the outbreak and led to the expansion of the Office of the Surgeon General, which monitored the spread of cholera throughout the Reconstruction South and the West. A number of nations sought to observe the spread of cholera and avert outbreaks, and they all did so in a similar way: those in power established surveillance methods to observe the spread of cholera among various populations.[15]

The outbreak of cholera in the mid-nineteenth century encouraged physicians to think more globally. Beginning in 1851, officials from various European countries met in Paris, which was considered the hub for diplomatic relations, to form the International Sanitary Commission (ISC). The ISC included representatives from France, Great Britain,

Russia, Austria, Sardinia, Tuscany, the Papal States, Naples, Turkey, Greece, Spain, and Portugal. Gavin Milroy, a leading British epidemiologist who had written about cholera in Jamaica in 1850–1851 (see Chapter 4), published a report about this gathering, and noted the absence of the United States. "The omission from such an inquiry of the United States, which could have afforded the most valuable information on some leading points, was, I think, much to be regretted."[16]

Members of the ISC combined their bureaucratic and military forces in order to observe cholera and other epidemic diseases beyond their national borders and to collectively monitor them throughout the world, especially at the eastern borders of Europe. Chief among their concerns was to identify the beginnings of an epidemic and enact the necessary quarantine measures to prevent it from entering Europe. According to Milroy, the ISC addressed three quarantine practices: 1. quarantine of observation, which required the detention of a vessel for a specific time, typically a few days, and involved observation of the ship's sanitary conditions and inspection of its ventilation capacities; 2. strict quarantine, which was a more prolonged detention that required passengers and crew to disembark and stay at a quarantine facility and cargo to be unloaded; and 3. suspected quarantine, which Milroy claimed the ISC resolved should be discontinued.[17] The nations finally agreed that quarantine measures should be adopted for three diseases—plague, yellow fever, and cholera—but there was controversy about the usefulness of quarantines for cholera. According to Milroy, Austria protested the most, based on its efforts to enact quarantine measures during the 1831–1832 cholera epidemic. Austria argued that quarantine was "useless" and even "disastrously mischievous." Great Britain, France, and Sardinia also opposed it, while Naples and the Papal States urged the necessity of quarantine. They claimed that Elba and other places in Italy "had been preserved from the pestilence by the adoption of strict segregation and the exclusion of all suspected arrivals." Portugal and Spain argued that until sanitary measures could be guaranteed on board merchant vessels and at ports, strict quarantine measures should continue. Russia had had mixed results with quarantine in previous cholera epidemics and wanted to wait for the results of additional investigation.[18]

The British opposition to quarantine as a general principle grew out of observations and reports that the government had been collecting for decades in Africa and the Caribbean. The debates about quarantine and contagion that animated British concerns about the arrival of the yellow-fever-infested ship the *Eclair* from Cape Verde had unfolded only a few years before, in 1844. Milroy himself had studied cholera in Jamaica and come out firmly against quarantine on the grounds that cholera was not contagious. The French and Austrians also argued against quarantine because they believed it was an outdated, ineffective way of preventing the spread of disease. Technological advancements, such as steamships and railroads, made it impossible, they claimed, to restrict the mass movement of people. The focus should instead be on making sure that Russia and other border countries took measures to keep cholera from entering Europe. Although a majority of the members voted to impose a quarantine of five days on arrivals from locations where cholera was found, the conference ultimately did not reach any conclusive outcomes; the final convention was signed by only three countries, two of which later withdrew.[19]

Despite the fact that the first International Sanitary Commission conference did not result in a resolution, the organization of the meeting furthered the development of epidemiology through the members' attempts to visualize the origin and movement of cholera. By coming together as a group, they attempted to gain a bird's-eye view of cholera's movement. That said, their prejudices tainted their efforts to track the disease. They theorized that cholera originated in less-developed places in the world. While they correctly identified India as the source of cholera, this conclusion rested on a prejudice that India's inferiority to Europe caused cholera to escalate.[20]

A second ISC conference was convened in 1859, with only diplomats in attendance. They did not want to engage the scientific arguments about disease transmission. During this conference, which lasted from April to August but again resulted in no signed convention, the delegates' efforts to track cholera's movement became increasingly prejudicial. Although they conceded that there was a risk of cholera being spread via military ships, they recommended that such ships be exempt from showing a clean bill of health at the discretion of the commander.[21]

By the time of the next ISC meeting, in Constantinople in 1866, a new cholera epidemic had spread to Europe and throughout many parts of the world, including the United States, the Caribbean, and parts of South America.[22] The epidemic had killed over ten thousand pilgrims in Mecca in 1865, followed by an outbreak in Egypt after surviving pilgrims arrived in the port of Suez. Members of the conference made paradoxical assumptions about the movement of people. On the one hand, they indicted people in the East for being sedentary and unlike modern Europeans, who traveled widely; but when Muslims crossed national borders as part of their pilgrimage to Mecca, the French and others condemned them. In fact, the French delegates proposed that Muslim pilgrims be quarantined before leaving the environs of Mecca. The proposal narrowly passed.[23] The 1866 conference led to a hardening of the distinction between the East and the West and prompted Russia and the Ottoman Empire to try to prove that they were trustworthy, modern partners by upholding their commitment to enforcing sanitary regulations and preventing cholera from entering Europe.[24]

Yet, unlike in previous meetings, scientists participated in this conference, and they gained more authority. Although some knew of John Snow's theory that cholera was transmitted through contaminated water, there was considerable disagreement about what caused cholera and how it spread. The delegates concluded that cholera could be communicated through air or water, that it could be transmitted by clothes or linens, and that although its ultimate cause was unknown, it originated in India.[25]

While the International Sanitary Commission served as the global alliance that monitored the movement of cholera, various countries, including France and the United States, launched their own reports on the outbreak of cholera in their respective countries and globally. In Britain, John Netten Radcliffe, secretary of the Epidemiological Society of London, was commissioned by the Privy Council to gather together materials from the foreign office to analyze the 1865–1866 cholera epidemic.[26] Radcliffe believed that previous studies of cholera epidemics had been incomplete and inaccurate. He argued that epidemiology, which was still in its early days of professional development, ought to follow the practices of meteorology, which offered a set of scientific principles for how to conduct investigations. Both fields, Radcliffe posited, "must equally rest on accurate data collected in a wide area of observation, and over periods of time more or less extended."[27] Britain, wrote Radcliffe, was in an excellent position to study cholera in different locations because "Great Britain possesses greater facilities in obtaining such information, from her wide spread relations and intercourse, than most other countries."[28]

Radcliffe's claim that British authorities had "greater facilities" refers to the way that the bureaucracy of imperialism and colonialism allowed for the gathering of information from around the world. Radcliffe requested that details on the outbreak and progress of cholera be communicated to him from British consuls. But this claim also refers to Britain's ability to obtain information from other nations. "Official and trustworthy reports of the dissemination of the epidemic in different kingdoms and states of the continent of Europe and in North and South America are but slowly appearing," he noted. Although it was taking time for these reports to arrive and be processed, it would have been difficult to obtain them at all, he wrote, without "the intervention of Government." The outbreak of cholera in the nineteenth century advanced international collaboration among nations and illustrates how nations depended on each other's reports to track cholera, to understand its pathology and etiology, and to see it.[29]

By the mid-nineteenth century, many governments throughout the world had begun to establish surveillance methods to observe the outbreak of epidemics by paying close attention to any sign of public health disruptions. As Radcliffe notes, no nation had reported evidence of slight health problems just before the cholera epidemic of 1865–1866 hit, as they had before the two previous epidemics. For the few doctors who did offer a semblance of evidence that could signal the coming of an epidemic, Radcliffe dismisses their findings. He notes that John Sutherland, the British physician and sanitarian, noticed a slight uptick in diarrheal disease in Malta during the six months preceding the epidemic, but points out that no such uptick had occurred in Gibraltar. Radcliffe acknowledges that in Egypt the cholera epidemic occurred during a period of "great privation," but cholera also attacked other communities that were not suffering from privation. "The truth is that the epidemic took Europe by surprise," Radcliffe concludes, "and so generally good was the public health that the first news of the appearance of cholera in Egypt excited no attention and scarcely any alarm. It was not until the disease had effected a lodgement in several parts of Europe that the notice of governments and of the public was fully aroused."[30]

Radcliffe traced the beginning of the cholera epidemic to two British ships that arrived in Jeddah in March 1865, having come from Singapore. The ships were full of mostly Javanese Muslim pilgrims who were headed to Mecca. Cholera appeared on the ships after they had stopped at Makalla, a port in the southern Arabian Peninsula (now Al-Mukalla, Yemen), and 145 people, both passengers and members of the crew, died. Other ships that arrived in Jeddah from India in March and April 1865 also reported cholera deaths. The captains of the ships from Singapore claimed that their passengers had contracted the disease in Makalla. Officials in Makalla denied that their port was infected, but Radcliffe noted that there had been reports of cholera at a number of nearby locations. Cholera, he concluded, was probably prevalent in southern and western Arabia before the arrival of the pilgrims, as well as along parts of the African coast.[31]

As evidence for the existence of cholera in Africa, he referred to the capture by the HMS *Penguin* of two vessels in the Gulf of Aden that were carrying enslaved people.[32] Many of the enslaved people died of cholera, as did two crew members of the *Penguin*. While his account does not give any details about the enslaved people, an article published in *The Illustrated London News* in 1867 does. It reports that there were about 216 enslaved people on board, "chiefly boys and girls, closely stowed, and all bound for Muscat, in the Persian Gulf." The children were "well-behaved and kind to each other, sharing everything which was given to them." Some of the youngest were "emaciated," but most were "healthy and in good condition." The "Arab captain" of the slave vessel "was dying from fever when he first came on board" but eventually rallied after treatment.[33] The children were transported to Aden, one of the few places where British vessels policing the region for illegal traffic of enslaved people could safely disembark captured enslaved Africans. Unlike on the west coast of Africa, where British officials could release captured Africans in Sierra Leone or Liberia, places

where Africans lived freely, British officials worried that Africans released on the east coast of Africa might be recaptured by slavers stalking
the Indian Ocean.[34] British policing vessels, as a result, sent captives to
Aden or Bombay (Mumbai), but many then died there.[35] Captured Africans, like Muslim pilgrims, made cholera visible to Radcliffe. The story
of their plight became part of a report that contained details about their
health that then provided evidence about cholera in Africa. As in other
periods and places, slavery produced a discourse that made infectious
disease visible.[36]

With access also to the history of this region, Radcliffe noted that
Bombay was "barely recovering from the severest outbreak of cholera
from which the island . . . had suffered for many years." He believed that
cholera could have spread from Bombay to Makalla ("not necessarily
carried by pilgrims") and from there to "other towns of the Arabian
coast as well as to the coast of Africa." Quoting the British health officer for Bombay, who reported that smallpox was often brought into
Bombay by people arriving from Arabia, Persia, and Africa, he concluded, "It is no fanciful theory that these coasts have a traffic in diseases as well as goods."[37]

Radcliffe regarded the Muslim pilgrimage as the cause for the spread
of the cholera epidemic in 1865, claiming that the pilgrims carried
cholera with them on their journeys back to their homelands. "Cholera
smouldered in Jedda from the time of the arrival of the infected ships,"
he reported. Even before the pilgrimage was complete, the disease broke
out "with great violence," and "its ravages became so alarming, that as
soon as the rites were completed the assemblage broke up in a panic,
and the different caravans hastened their route homewards, the disease
following their steps."[38] Radcliffe frames his description of the cholera
epidemic in terms of fear and even violence, which does very little to
illuminate the pathophysiology of cholera but illustrates the abrupt, haz-
ardous outbreak of the epidemic.

In London, Radcliffe used military reports to pinpoint the movement
of the various vessels and the massive mortality that occurred. After finishing their rituals in Mecca, the pilgrims traveled roughly fifty miles
overland to Jeddah, where they boarded ships to Suez, Egypt. He recorded with precise detail the dates that ships arrived and departed in

an effort to gain a better sense of the movement of the epidemic. On the first British ship to leave Jeddah, several of the 1,500 pilgrims on board died along the way from "unknown causes" and their "bodies had been cast into the sea." A few days after the ship arrived in Suez, on May 19, 1865, the captain and his wife were "both attacked with cholera." While the Muslims who died on the voyage likely succumbed to cholera, Radcliffe did not identify a cause of death. On one level, it is not surprising that the report that flowed back to the British metropole did not contain an accurate description of the cause of death. The chaos, fear, despair, anguish, suffering, and stench that accompanied the deaths probably offered little opportunity for any medical official to properly diagnose or even treat these people. However, the captain and his wife were diagnosed with cholera in Suez. British medical authorities likely did not hesitate to talk to the captain and his wife about their symptoms or even to touch their bodies, which led them to be able to diagnose them. But the Muslims aboard the ship were not diagnosed. Instead, they died and were thrown overboard, their deaths classified simply as "unknown causes."

Radcliffe further tracks the movement of the many thousands of Muslim pilgrims who landed in Suez between May 19 and June 1. Some returned to their homes in Egypt. Others, with support from the Egyptian government, traveled by train to Alexandria, where they camped outside the city waiting for ships to take them to their next destination. On June 2, the "first recognized" case of cholera appeared in Alexandria in a person who had been in contact with the pilgrims. More cases emerged in the next few days. Until June 12, according to the French physician to the Suez Canal Company, these cases were restricted to people who were in direct contact with the pilgrims, but by the end of July the epidemic had spread throughout the country, causing over sixty thousand deaths within three months.[39]

Drawing on reports from British officials stationed throughout the Mediterranean, Radcliffe detailed how the cholera epidemic traveled from North Africa to Malta, which served as a major hub for ships that made their way from Alexandria to other Mediterranean ports. The first reported cholera case of the 1865 epidemic in Malta occurred on June 20, and Radcliffe attempted to track the arrival of ships in Malta in order to map the trajectory of the epidemic. The Muslim pilgrims began arriving in Malta on May 31, when a ship bound for Tunisia arrived from Alexandria. Of the 200 pilgrims on board, 61 remained in Malta. According to a report by Antonio Ghio, the chief physician at the Malta quarantine facility, or lazaretto, fourteen steamers had traveled from Alexandria to Malta between May 31 and June 14 with a total of 426 pilgrims out of 845 passengers. None were quarantined.

On June 14, however, news reached Malta that cholera had broken out in Alexandria, and Maltese authorities immediately imposed a quarantine of seven days on any vessel arriving from that city. That same evening, a vessel arrived with eleven pilgrims aboard; the captain reported that one pilgrim had died from a bowel complaint during the voyage. On June 20 another vessel from Alexandria arrived in Malta and reported that during the voyage one crew member and one passenger had died from cholera. On the same day, Maltese authorities identified the first case of cholera on the small island where the lazaretto was located. At the time 288 people were being detained, and the distance between the first reported case and the lazaretto was 662 feet.[40]

Radcliffe depended upon the movement of Muslim pilgrims in order to visualize the spread of cholera across a broad geography. On July 21, he reported, a sailor from Malta had visited his sisters in Gozo, a neighboring island, and was "seized" that night by cholera but recovered. Three days later, there were four more cases—two of the sailor's sisters, a relative, and a resident of the village. Radcliffe's report tracks the disease as it moved to Greece, Turkey, and Beirut, where it "smouldered" in the city for the first two weeks of July. He then tracked Persian pilgrims who traveled from Beirut through Aleppo, Syria, on their way home, describing their sanitary condition as appalling: "part of [the caravan's] baggage consisted of the corpses of fellow pilgrims who had died on the journey." The first case of cholera appeared in Aleppo on August 15. Within three months, 7,000 were dead out of roughly 90,000 inhabitants. Radcliffe continued to follow the pilgrims on their return home, noting that "cholera accompanied the caravan . . . along its whole route to the Euphrates."[41]

As the epidemic made its way through the Middle East in July, a French merchant who had traveled to Valencia from Alexandria via

Marseilles died, as did others in the house where he was staying. "This was the starting point of the epidemic in Spain," noted Radcliffe. On the same day, the disease appeared among residents of Ancona, Italy. A few days earlier, a woman who had arrived from Alexandria had been placed in the lazaretto in Ancona for six days. The next day, after she left for Pistoia, she came down with cholera and died. Radcliffe followed other fragments of evidence to track the movement of cholera to and through Europe. He pieced together details about the presence of the disease in Gibraltar and in Marseilles, where the first official victim, on July 23, was a nurse taking care of an infant whose father was a crew member of a steam vessel. But Radcliffe suggested that the disease had already reached Marseilles; starting from the second week of June, many pilgrims returning from Mecca had passed through the city on their way home.[42] As the epidemic spread from the Middle East to the Mediterranean and beyond, Radcliffe expanded his gaze to Europe and then across the Atlantic Ocean to the United States. The first reported case in Britain appeared in Southampton, England, on September 17.

While most of Radcliffe's analysis very clearly follows the spread of cholera from one location to the next by tracking specific vessels, the next section of the report covers the broadest geography but is given the shortest description. "It would take up too much time," he explained, "to trace its development along the shores of the Euxine and Sea of Marmora, following the track of infected vessels from Constantinople; its ascent of the Danube, and penetration of Russia from the Black Sea provinces." He only briefly mentions the spread of the epidemic from Ukraine to Germany, and from Europe to New York City and Guadeloupe.[43]

His failure to describe the epidemic moving west through Europe and to the United States and the Caribbean could have resulted from the limitations of the military bureaucracy and gaps in the burgeoning alliance among nations throughout the world to both compile and share reports. The British Empire deployed medical officers in various stations throughout the globe who sent reports to Radcliffe, which enabled him to see cholera moving from the Middle East to the Mediterranean. Radcliffe tracked the early stages of the epidemic farther east from Syria. "I have followed the track of the epidemic to the north and west and from the east coast of Syria. Now I must show its course in another direction." He described the disease breaking out along the Euphrates River, through "Kourra, Suk-el-suk, Samava, Divamieh, Iman-Ali, and Kerbellah," and eventually reaching Baghdad in September. "Along this route," he maintained, "the development of the epidemic coincided with the passage of the Persian and Central Asia pilgrims returning from Mecca by sea, ascending the Gulf of Persia and the river Euphrates."[44]

Moving to Africa, Radcliffe claimed that "Takruri pilgrims" (West African Muslims) spread cholera to Massawa, Abyssinia (now Mitsiwa, Eritrea), on their way back from Mecca.[45] As evidence, Radcliffe cited "official correspondence on Abyssinia" along with *The Story of the Captives,* a popular narrative written by a physician, Henry Blanc, who was an assistant surgeon in the Bombay Army. Blanc had left India and traveled to Abyssinia (now Ethiopia and Eritrea) with a British delegation to rescue British missionaries who had been taken hostage by the emperor of Abyssinia. Three months after Blanc arrived in Abyssinia, he was imprisoned. Blanc's imprisonment intensified an international crisis between Abyssinia and England, making headlines in British newspapers. Blanc was eventually released and wrote a bestselling account of the ordeal, which included details about the transmission of cholera to Africa by Muslims returning home from the hajj. Radcliffe assumed that his readership was familiar with Blanc's narrative.[46]

Although a story that enthralled English readers with its harrowing details about imprisonment in Africa may appear on the surface as a suspect or, at best, an imprecise source for Radcliffe to use, despite the fact that it was authored by a physician, the book's narrative mode was not radically different from that of the reports that Radcliffe received from British and European medical inspectors and bureaucrats. Many physicians who tracked cholera wrote in narrative form. Like their American counterparts who described disease outbreaks in the Civil War South and West, they often described the environment where cholera erupted, the conditions in the region, and the characteristics of the people. In their efforts to understand the spread of the epidemic, they crafted narratives that explained how disease traveled from one location to the next.

Epidemiology in the 1860s did not universally rely on statistics to document the spread of disease. It relied on documented accounts that provided a story that explained the cause, spread, and prevention of infectious disease. Radcliffe amassed these narrative reports and consolidated them into an even larger story that offered one of the earliest accounts of a pandemic; many previous studies on medieval epidemics or other outbreaks had been limited to specific local regions.[47] The bureaucracy created by the British empire enabled Radcliffe to see across a much more capacious terrain. According to Radcliffe, the 1865–1866 epidemic spread more quickly than the previous epidemics of the nineteenth century because of increased global traffic: "This unparalleled rapidity of progress finds no explanation in any peculiar virulence of the disease, but solely, as far as I can see, in the increased facilities and greater rapidity of traffic between different countries."[48] The reports enabled Radcliffe to "see" and thereby theorize about the spread of the epidemic.

✦ ✦ ✦

The US Surgeon General's Office began the process of collecting observations from military doctors stationed throughout the South and West six months after the cholera pandemic struck the United States in 1866. During the Civil War the US Sanitary Commission had mobilized the Surgeon General's Office, which had mainly overseen the health of merchant seamen, to take a more active role in monitoring the health of the country.[49] The expansion of the Surgeon General's Office also grew out of the Medical Division of the Freedmen's Bureau, an agency created by the War Department to respond to the alarming outbreak of disease that disproportionately affected newly freed people in the Reconstruction South. The Medical Division established the first system of medical care in the United States. It built 40 hospitals, employed over 120 physicians, and provided medical care to roughly one million formerly enslaved people. By depending upon military physicians to provide the federal government with reports on health conditions within the postwar South, the Freedmen's Medical Division led to the expansion of federal authority. Doctors filed reports that documented the number of sick people in hospitals, measured mortality rates, and depicted the health conditions in the region. In 1865–1866, when a smallpox epidemic ravaged mostly the Black population, doctors and other federal agents described in unsettling detail their inability to halt the outbreak. They claimed that there were not enough physicians to treat the sick and that they lacked resources to build quarantine facilities. Federal agents claimed that Black people were innately vulnerable to infectious disease and that their supposed immorality and indolence made them especially susceptible to smallpox.[50]

When cholera reached the South in 1866, in the midst of the smallpox epidemic, all of the reasons that federal agents provided for not being able to prevent smallpox disappeared. Cholera was threatening the lives of both white and Black people, whereas smallpox, it was claimed, primarily infected Black people.[51] Using the Freedmen's Bureau as a blueprint, the Surgeon General's Office expanded its power and authority. It drew on military doctors' reports to track the epidemic as it spread through the US Army and to offer guidelines on prevention, which mostly included sanitary precautions.

The Surgeon General's Office published a comprehensive report on the cholera epidemic "for the information and guidance of medical officers." The report, addressed to Joseph K. Barnes, surgeon general of the US Army, was put together by Joseph J. Woodward, assistant surgeon in the army, who was employed by the Surgeon General's Office and also headed the Medical Section of the Army Medical Museum. Woodward collected information from army physicians and put together a report on the 1866 cholera epidemic in the military. (The following year he produced a similar report for 1867 that also covered yellow fever.) In the report he noted that despite the fact that the number of cases of cholera in the army was "not very great," information related to the spread of the disease could reveal crucial insights about "the question of quarantine" and so "appears well worthy of the attention of all interested in problems of public hygiene."[52] In the same way that Joseph Jones collected Confederate doctors' observations about the effects of smallpox vaccination (see Chapter 7), the Surgeon General's Office recognized the value of doctors' observations regarding the behavior of cholera.

By collecting the reports of doctors in the army, Woodward was able to map the spread of cholera from one location to the next and produce the largest study of the spread of infectious disease in the United States to date.[53] Prior to this, municipalities and state governments had limited means to observe the spread of infectious disease beyond their borders.[54]

For example, in 1832, fifteen members of the New York Medical Society formed a special commission to study an earlier cholera outbreak, but their report was limited to New York State.[55] In fact, the outbreak of cholera in 1866 coincided with the expansion of the federal government, which grew out of Reconstruction and gave way to an extensive military bureaucracy that allowed federal authorities to observe how cholera spread throughout the entire nation and even into the Caribbean and Central America.

Woodward, in the surgeon general's report, noted that the first cholera case in the army was identified at a fort on Governor's Island in New York Harbor. A recruit from Minneapolis first showed signs of infection on July 3, 1866. A few hours after he was admitted to the hospital, another recruit also presented symptoms of cholera. The disease then spread to neighboring Hart's Island. Due to the severity of the epidemic there, the troops were moved to another location, David's Island. On July 19, a soldier in Boston who had arrived that morning from Hart's Island died of cholera, but no other cases appeared in Boston.[56]

The bureaucratic records enabled the Surgeon General's Office to document how and where cholera did or did not spread. A steamship named the *San Salvador,* which included about 140 crew and passengers, left New York on July 14. It stopped at Governor's Island to pick up 476 recruits for the 7th US Infantry. On the second day of the voyage to Florida, cholera broke out among the recruits, who were crowded into a space between decks. By the time the ship made it to Savannah four days later, three men had died and twenty-five were sick. The *San Salvador* was placed under quarantine. It docked at neighboring Tybee Island, where the military had created a makeshift hospital. For the next three weeks the remaining recruits had to survive on the island. During that period, "cholera continued to prevail on the island"; 116 people died of cholera, and 202 became ill. A number of soldiers attempted to escape, and eighteen of them were later found dead. All of the ten white people living on the island became ill; five died, and one was later found dead "somewhere in the interior of Georgia." No cases of cholera were reported among those who quarantined on the ship or among troops in Savannah.[57]

In the same way that Radcliffe followed Muslim pilgrims to visualize the spread of cholera, the Surgeon General's Office followed recruits. Collecting reports from medical officials across the nation enabled Woodward to observe how the movement of troops around the country led to the spread of cholera. He traced the outbreak among troops in New Orleans to recruits who traveled from New York to Louisiana on the *Herman Livingston* steamship. Two soldiers had died of cholera along the way, and a few ailing men were removed from the ship after it arrived on July 16. A couple of days later, most of the troops headed to Galveston, Texas.

On July 22, after the New York–based crew left for Texas, a member of Company G, 6th US Cavalry, stationed in New Orleans, came down with cholera. It was unclear if he had had contact with the New York recruits. Then a cholera case emerged among one of the New York recruits still in Louisiana, and a number of additional cases occurred among the cavalry. On July 25, a soldier of the "81st colored infantry at the Louisiana cotton press" contracted cholera and died the next day. The following day another Black soldier who had been on guard duty at "Bull Head's stables, near the levee," was brought into the camp ailing and died the day after. The disease then "spread rapidly through the regiment." It was already present among civilians in New Orleans, and some of the first soldiers to be attacked were "brought in a state of collapse from hovels in the city."[58] New Orleans city officials did not put a health ordinance into effect until after cholera had already infected a number of people, so it was not possible for municipal authorities to pinpoint the early cholera cases in the way the military could.[59]

From New Orleans, cholera accompanied the troops that traveled to Texas and then made its next "appearances" in Virginia, Georgia, and Kentucky. The map that military authorities constructed had less to do with the actual spread of cholera than with the bureaucracy that allowed them to see it. There may have been other cases in the military that were not reported. There were also instances when the military bureaucracy captured details about the civilian population, as evidenced by the information collected about the epidemic among the civilians in Louisiana. The Surgeon General's Office was only able to follow the movement of the disease based on the reports it received. As Woodward explains about the epidemic that traveled from New Orleans up the Mississippi River, "though the whole chain of evidence is not complete, yet there are a sufficient number of known cases of the transfer of the epidemic from one post to another in this region to put this view of the whole movement beyond reasonable doubt."[60]

In order to visualize the spread of cholera, military doctors relied on a number of terms and expressions that described the presence and movement of the disease. Both military and civilian officials who filed reports on cholera often used the verb "appear" to describe the presence of cholera. The term "appear" also became a way to mark a particular region that had people infected with cholera. For example, Woodward wrote that cholera "appeared" in the city of Richmond among its citizens.[61] While the term "appeared" served as a logic to explain the presence of cholera often not under military surveillance, many other officers continually remarked that certain individuals "carried" the epidemic. Woodward, for example, notes that cholera "was carried [to troops in Austin] by recruits who arrived by way of Indianola" and was "carried" to Little Rock by steamboats.[62] Officials also used familiar idioms to describe the presence of cholera. For instance, in describing cholera in Shreveport, Louisiana, in the 80th US "colored troops," Woodward noted that cholera cases "had been reported on plantations below."[63]

✦   ✦   ✦

Military officials also recorded the effect of a political uprising on the spread of cholera through troops in Mississippi and Louisiana. In 1866, Republicans were the champions of freed slaves, and white Southern Democrats were their adversaries. Two years earlier, Republicans had begun the process of extending rights to Black men, but Democrats became angered by these efforts and passed legislation to prevent Black men from gaining the right to vote in Louisiana. Republicans, in turn, called for the Louisiana Constitutional Convention to meet in order to extend suffrage to Black men and stop the Democrats from passing laws that curtailed Black people's newly won freedoms. Meanwhile, two days before the convention, Black military veterans had organized a meeting in New Orleans on July 27, 1866, to support Republican efforts. They listened to speeches given by prominent abolitionists and became emboldened to fight for suffrage based on their military service. When the convention took place on July 30, they marched with a band playing by their side to the Mechanics Institute. White Democrats, many of whom had served in the Confederate Army, became infuriated not only because of the protest but because they were opposed to extending suffrage to Black people. They, along with the police, began brutally attacking the Black veterans. Accounts of the massacre, which killed nearly fifty people, tell of police bludgeoning the heads of Black men who had already been kicked to the ground, white sex workers calling for the death of Black veterans, and Black men's bodies being left on the street. Prisoners were later charged with collecting the bodies and placing them in carts to be transported to a burial ground, but some of the men were later found to be alive.[64]

The military deployed Black troops from the surrounding area to quell the unrest. When the troops eventually returned to forts along the Mississippi River south of New Orleans, they began to present symptoms of cholera. Military doctors believed that the troops had become infected with cholera while on duty during the uprising.[65] Violence served as a coefficient to visualize the spread of violence. Cholera was a dangerous menacing force that threatened people's lives throughout the world in the nineteenth century, but its most conspicuous manifestation often emerged among subjugated populations and was narrated by employing the language of violence.

For government and military authorities in Washington, DC, the references to people of color as carriers—which is analogous to Muslims returning from the hajj—helped them to see the epidemic spreading. Detailing the spread of cholera in Vicksburg, for instance, Woodward's report noted that the first case among white troops happened on August 22, 1866, but "the colored barber had died of the disease the day before," possibly implying that the barber was the cause of cholera to spread among the white regiment.[66] Additionally, Woodward's report blamed a cholera outbreak among a detachment of the 17th US Infantry near San Antonio on "two Mexican teamsters who came from San Antonio, stopped for the night near the camp, and died of the disease."[67] Similar to the cases of laundresses in Malta and hospital attendants in

India in the first half of the nineteenth century, authorities here were using subjugated populations to help visualize and track the spread of infectious disease.

Woodward's report documented the spread of the disease in both white and Black troops. After tracking the epidemic from New York to Louisiana, Woodward described its movement to other parts of the country. He noted its presence in Kentucky, Virginia, Arkansas, New Mexico, and Kansas. He attempted to connect some of these outbreaks to Louisiana, but he admitted that he was not able to identify every link in the chain. He nonetheless concluded his report by mentioning a cholera outbreak on a ship in the San Juan River in Nicaragua, among a group of recruits on their way from New York to San Francisco.[68]

The accumulation of the military doctors' reports helped to solidify knowledge about the pathology of cholera and to explore the cause of the epidemic.[69] A military doctor based at the quarantine station in Tybee Island, Georgia, for example, reported that the men who were "suddenly seized" by severe symptoms invariably died soon thereafter. One common symptom was "rice-water vomiting."[70] Other military physicians referred to "rice-water stools"—voluminous, watery diarrhea with flecks of mucous that results from the way the disease attacks the small intestine.[71] A military doctor who visited the hospital on Hart's Island concluded that "the disease was undoubtedly cholera Asiatica" and then listed nearly two dozen symptoms characteristic of the "malignant form," including "vomiting and purging of rice water."[72]

In the era before the germ theory of disease, physicians often needed to observe for themselves how a disease unfolded on the ground before they adopted a theory of what caused an outbreak. For example, the army surgeon John Vansant, who was stationed in Little Rock, Arkansas, didn't know the cause of the outbreak that occurred at the arsenal and wondered whether there was any use studying conditions in the place where it broke out. He was unsure whether cholera, "like other infectious diseases, has a period of incubation." If it did, investigating the place where it erupted might be misguided, since it could have been contracted in another region. Vansant considered the role of atmospheric conditions, noting that the "oxygen of the air" at the time had "singular activity," promoting rust despite no increase in moisture. But this did not appear to lead anywhere; he concluded, "My observations of this epidemic point to nothing which I can imagine might probably be the local cause."[73]

Several of the doctors did mention the quality of the water as a factor, however; and in his report, Woodward noted that "the importance of the character of the drinking water used during epidemics of cholera had attracted attention in Europe" and that the English registrar general had found that "the prevalence of the disease in the several districts of London bore a direct proportion to the amount of the organic impurities of the water."[74] When cholera first appeared among the troops in New York Harbor, samples of the drinking water were sent to the Surgeon General's Office for analysis by Benjamin F. Craig, who directed a chemical laboratory at the Smithsonian Institution. Trained as a physician, Craig studied chemistry and physical science in London and Paris and served as a professor of chemistry at Georgetown Medical College for three years before becoming head of the chemical laboratory at the Smithsonian in 1858.[75] Having trained in London, Craig likely maintained an international professional network, either through personal correspondence or access to medical journals, that may have alerted him to John Snow's waterborne theory of cholera.

Craig observed "considerable quantity of organic impurity" in the water samples, especially from two of the sites. He recommended that the water be purified with permanganate of potash (potassium permanganate) before being used for drinking and supplied instructions for how to do so.[76] According to Woodward, "this recommendation, so far as known, was not acted upon," and there was not even much attempt made to find pure water sources, other than in New Orleans.[77]

The reason that military authorities did not follow Craig's recommendation for purifying water in 1866 remains unknown, but there are some possible explanations. The Surgeon General's Office may not have properly conveyed this information to all physicians. In the immediate aftermath of the Civil War, the government was just beginning to stretch its bureaucracy across the country and to send out messages and circulars. Given the difficulty of relaying information, it may have been even more difficult to send permanganate. Many physicians bemoaned the lack of even basic supplies, such as tents, blankets, and food.[78] Even if they had been able to obtain the chemical, it was not presented as a magic bullet to prevent cholera but rather as one of many disinfectants to ward off the spread of disease. A doctor stationed at a quarantine station on Tybee Island near Savannah, Georgia, noted permanganate among his supplies but appears to have used it as a general disinfectant, not to purify water. He blamed the severe outbreak there on a diet of green vegetables, lack of ventilation in the makeshift tent hospital, a shortage of physicians and supplies, and the fact that the "grounds were . . . unpoliced," which meant a lack of sanitary enforcement.[79] As the doctor's testimony reveals, physicians were often more invested in cleaning as a form of prevention than in identifying sources of contaminated water. They were also more focused on treating patients than in prevention.[80]

Despite the failure of many physicians to purify water during the cholera outbreak, the surgeon general's report recorded all the preventive measures that physicians did undertake—changing diet, increasing ventilation, moving the camp to a different location, disinfecting latrines, instituting quarantines, burning clothing, and so on. By documenting these activities, the surgeon general promoted sanitation as a key response to the outbreak of cholera. While in retrospect the most important contribution of the report was that it promoted use of permanganate as a water purification measure, its focus on sanitation was profoundly significant for 1866. Just a few short years earlier, when the Civil War first began, the surgeon general failed to respond to the medical catastrophes that the war generated; this led to a civilian group establishing the USSC, devoted entirely to addressing sanitary issues. The 1866 surgeon general's report continued the USSC's work by documenting doctors' efforts to promote a sanitary environment.

Before the creation of local, state, or federal public health authorities, the military advanced the development of epidemiology by drawing on the expansive knowledge gained from monitoring heath conditions among troops across a broad geography. Relying on military bureaucracy, the federal government developed a bird's-eye view of the spread of cholera through the US Army. Before the widespread use of maps, the Surgeon General's Office relied on detailed narrative reports to map the spread of the epidemic across the country. Military authorities received reports from doctors stationed throughout the country that enabled the federal government to see cholera in a way that local and state governments could not.

Woodward's comprehensive report on the 1866 cholera epidemic in the army marks a turning point in the history of epidemiology in the United States; the federal government collated a set of practices and ideas from across the country that aimed to prevent a future epidemic from spreading. When a cholera epidemic threatened the army in 1867, Woodward sent a copy of his 1866 report, also known as Circular No. 5, to "medical officers" throughout the army. Military bureaucracy facilitated the spread of knowledge about the epidemic, which alerted physicians and officials to the threat and allowed them to respond quickly to the first reported case before it snowballed. For example, several doctors in 1867 reported using permanganate to purify water for troops, which likely mitigated the spread of cholera.[81] Woodward attributed the low rate of cholera deaths in 1867 to the fact that physicians read the circular, which argued that the movement of troops, particularly of recruits, had caused the widespread outbreak in 1866. Woodward noted that the infection and mortality rates were higher in 1866—2,724 cases and 1,217 deaths—than in 1867, when only 230 cholera deaths were recorded.[82] As Woodward explains, the surgeon general's circular had recommended the use of quarantines, and "the [quarantine] measures thus adopted, in conjunction with the hygienic precautions directed in the same circular, undoubtedly saved many lives in the army, for the total number of deaths from cholera during 1867 was but 230, and it cannot be claimed that the disease in itself was less virulent during 1867, for the proportion of deaths to the total number of cases was 1 death to 2.19 cases, while during 1866 it was 1 to 2.22."[83]

✦   ✦   ✦

The postwar period had transformed the United States into precincts populated by armies and physicians who could supply the information that allowed Woodward to make this comparison between 1866 and 1867. Ideas about the cause, spread, and prevention of cholera circulated as a direct result of this deployment of troops as well as the displacement of Native Americans and the emancipation of freedpeople.

Knowledge about cholera—how it was written about by military officials on the ground, sent to authorities in Washington, DC, and then published as a major report—followed a pattern similar to those that developed in the British Empire and the Confederacy about infectious disease. War—like slavery and colonialism—produced social arrangements that enabled medical authorities to investigate the cause, spread, and prevention of infectious disease.

The Surgeon General's Office, McClellan, Radcliffe, and the International Sanitary Commission advanced the field of epidemiology by emphasizing the need to write reports to visualize the spread of cholera. Whereas later generations of epidemiologists would turn to actual maps to track the course of epidemics, the founding generation, with a few exceptions, turned to words and reports disseminated through bureaucratic means to see cholera's movement around the world in 1865–1866.[84] Within popular and even some scholarly circles, John Snow's famous maps of cholera transmission in a London neighborhood have become iconic illustrations of the development of epidemiology.[85] The military and medical narratives about the 1865–1866 pandemic provided a bird's-eye view of the epidemic, which advanced the development of epidemiology. The pandemic facilitated the production of narrative reports that enabled physicians across the world to observe outbreaks and to embolden institutions—the International Sanitary Commission and the US Surgeon General's Office—to develop methods to prevent pandemics from spreading.

Conclusion

The Roots of Epidemiology

B ETWEEN 1756 AND 1866, the medical community depended upon various populations around the world in order to advance the field of epidemiology. Medical authorities relied on enslaved and colonized people as well as soldiers and Muslim pilgrims to test theories and provide evidence to buttress arguments. It was not uncommon for doctors to rely on case studies during this period, but until now no one has systematically delineated the extent to which the medical profession depended upon mostly anonymous people, many of whom were enslaved or colonized, to understand infectious disease. The evidence that appears in this book represents only a small fraction of a much larger practice and pattern.

By highlighting some of these cases, *Maladies of Empire* shows how military and colonial bureaucracy facilitated the investigation of epidemic outbreaks among captive populations. Other scholars have described how, in other periods and places, medical, governmental, and religious authorities produced studies about dispossessed populations that flowed back to the metropole, but this book reveals the prevalence of this practice and traces its contribution to the development of epidemiology.[1] Doctors' correspondence and reports often articulated the first draft of a theory or insight that later evolved into a scientific principle. Working on slave ships, in colonial regimes, and on battlefields enabled medical professionals to become experts, but the people, places, and circumstances that informed their ideas were often not recorded. This book explains how the built environments that developed as a direct result of colonialism, slavery, and war made certain populations available for study and created the context for these new theories to develop. The use of slave ships, which helped confirm the benefit of ventilation systems and the importance of fresh air, illustrates how science made use of the violent and brutal conditions of the slave holds to visualize something as fundamental as the need for oxygen.

This book further underscores that the study of infectious disease cannot be separated from larger social transformations. Scholars have recently spilt a great deal of ink explaining how slavery contributed to the formation of modern capitalism.[2] Sven Beckert and Seth Rockman have argued that "American slavery is necessarily imprinted on the DNA of American capitalism."[3] This book follows in the spirit of that claim by showing how slavery is imprinted on the DNA of epidemiology. Slavery was designed as an economic institution, but we have seen that it contributed to the advancement of medical ideas and public health practices. Medical authorities exploited plantations and slave ships to observe the spread of infectious disease and to harvest vaccine matter. They used military camps full of sick soldiers and prisoners of war to create a comparison and context for sanitary improvements in cities. They traveled throughout the world to observe quarantine practices from the Middle East to the American West.

While scholars have often examined colonialism, slavery, and war as separate entities, this book has shown how medical professionals used these historical transformations to gather evidence about medicine and health. They then used this evidence interchangeably. The story of the British prisoners of war dying of suffocation in the crowded jail cell in India, for example, transpired a few years before Trotter's investigation of fresh air on the slave ship. Both cases provided the same evidence to doctors: that crowded conditions were unhealthy because they reduced the amount of fresh air. Doctors engaged in a similar practice of investigating epidemics among populations on slave ships, battlefields, and throughout the British Empire.

As this book shows, many modern epidemiological practices grew in part out of observing, treating, and preventing disease among captive populations produced by colonialism, slavery, and war. The medical personnel who studied epidemics often developed theories that are no longer considered correct. Their contribution, however, is less in the actual content of their theories than in the methods they developed. Their efforts to identify the cause of an epidemic, track its spread, record symptoms, develop a bird's-eye view of medical conditions in a region, and prescribe prophylactic measures to government authorities contributed to the foundational ideas of contemporary epidemiology and public health. Gavin Milroy's efforts to encourage government authorities to institute measures to stop the spread of cholera in Jamaica among a mostly Black population is an example of a public health innovation that can be traced to a location established by slavery and imperialism.

Military and colonial records, in turn, provide compelling and surprising new evidence. Epidemic outbreaks generated a proliferation of reports by a range of officials who investigated the spread of infectious disease. In each context, military and colonial officials were confronted with epidemic outbreaks that turned them into investigators. They collected information, wrote detailed notes, studied the landscape, and interviewed people who were sick or had witnessed an epidemic. This process created a trove of medical records that facilitated the development of epidemiology as a field.

By bringing slavery, imperialism, and war together into one study, *Maladies of Empire* shows that most doctors at the time thought of infectious disease primarily in terms of social and environmental factors rather than racial difference. Although many medical professionals subscribed to ideas about racial inferiority, they did not simply evoke that discourse in order to explain differences in the incidence of infectious disease. The medical discourse about racial inferiority emerged at specific times in specific places for specific purposes.[4] Much to my surprise, I discovered during my research that doctors, especially those in the British Empire, investigated the environmental and social factors that led to the spread of infectious disease rather than claiming that racial identity was the major cause. While both Florence Nightingale and Milroy certainly made derogatory claims about nonwhite racial groups, their main focus was on sanitary conditions. They collected copious amounts of information, documented their detailed observations, and analyzed the social landscape to support their theories that unsanitary conditions led to disease outbreaks. This book attends to the larger power structures of slavery and colonialism that enabled medical professionals to study these populations in the first place and reveals how colonialism, slavery, and war created built environments that led to the spread of infectious disease and provided the opportunity for doctors to try to explain these epidemics.

Although there were doctors during this period whose conclusions led them to support ideas about racial inferiority, *Maladies of Empire* highlights a different cast of physicians who transcended that framework and instead contributed to the development of methods to investigate how disease spreads within populations.[5] This book focuses on a group of doctors who studied large groups in an effort to visualize the spread of infectious disease. For the doctors highlighted in this book, the intellectual shift from the individual to the group resulted from the international slave trade, the expansion of colonialism, warfare, and the resulting population migrations that followed all of these. These larger social transformations created the impetus for these doctors to launch investigations into the environments in which the patients lived.

During the US Civil War, scientific racism became a metric for studying infectious disease. By the end of the Crimean War, Nightingale's emphasis on statistical evidence had pushed these studies toward quantitative methods, but when US doctors began to follow in her footsteps during the Civil War, they insisted on using racial identity as a category of analysis. Although the US Sanitary Commission was founded on principles of safeguarding the public health by encouraging adherence to sanitary principles, Union doctors groped, measured, and analyzed Black soldiers and in so doing infused racial ideology into the study of public health. After the war ended, they drew on Confederate doctors' medical reports to support their claims. US physicians validated race as a biological category. Today, racial identity remains a key metric used by public health authorities and epidemiologists to understand the spread of infectious disease. Even though slave-holders—not scientists—invented this idea, which has its institutional origins in the US Sanitary Commission, the field of public health and epidemiology remains steadfastly committed to it.[6]

While Northern doctors promulgated the significance of racial identity and deviated from the way British medical authorities were studying disease, Confederate doctors framed their studies more in line with their contemporaries across the Atlantic. At the end of a report on failed smallpox vaccinations, for example, the Confederate doctor Joseph Jones cites a number of European and British studies, including Trotter's study of a slave ship, in order to substantiate his argument. Jones understood infectious disease within a global context. As this book shows, physicians used the same examples, produced by colonialism, slavery, and war, to frame their arguments about the spread of infectious disease among large groups.

Although references to these case studies appear in treatises as sterile, seemingly objective facts, physicians often developed them in the context of a terrifying epidemic. There was a degree of contingency involved in the way medical authorities developed new understandings of infectious disease, but the bureaucracies that colonialism, slavery, and war created helped doctors craft narratives that transcended the chaos of the outbreak and enabled them to develop a bird's-eye view of how infectious disease spread. With this perspective, medical professionals created seemingly rational theories about the spread of infectious disease.

This bureaucratic process provided the foundation for contemporary practices of data collection and medical surveillance. Many members of the Epidemiological Society of London developed methods for tracking the spread of infectious disease that rose above the suffering and death the epidemics created. James McWilliam's meticulous interviews of the largely colonized population in Boa Vista, Cape Verde, provided evidence of the value of first-person testimony to understand the cause, spread, and prevention of epidemics and to offer specific details about the incubation period and the actual symptoms of a disease. Long before McWilliam arrived with his notebook to collect information, the local inhabitants had traced the movement of the epidemic across the island. This shows how ordinary people are often the first contact tracers. When the 1865–1866 cholera pandemic spread throughout the world, medical authorities from the Middle East to the United States surveilled the movement of Muslim pilgrims, Black and white troops, Native Americans, and other populations. Military bureaucracy provided these officials with the information to map the pandemic from one population and place to another.

Since the data collection and medical surveillance highlighted here were enabled by conditions created by colonialism, slavery and war, *Maladies of Empire* reveals how violence contributed to the foundation of epidemiology. The people whom McWilliam interviewed had little choice whether to cooperate. The soldiers and Native Americans whom military doctors monitored had to put up with the presence of military officials watching their every move, monitoring their health to see if they showed any symptoms of cholera. The study of infectious disease cannot simply trace its genealogy to biology, pathophysiology, or even urban populations; its genealogy includes military occupation, imbalances of power, and violence.[7]

For this reason, my aim in this book has been to shift the discussion in the history of medicine away from medical authorities to the people who made their theories visible. The people who became sick in a military camp or who were crowded in the bottom of a slave ship or who managed to escape an infection became the evidence that enabled doctors to visualize the spread of disease. Many of the people in the book appear only as fragments because of the uneven way that details about their identities entered the historical record. Many doctors discussed in the first few chapters did include the names of some of the people they studied. However, medical authorities' interest in recording names began to change by the time of the Crimean War and the Civil War. Nightingale's emphasis on statistics translated injured soldiers into numbers. The rise of epidemiology, with its focus on quantitative data, only further elided the human shape of the case studies and instead drew on information that could be quantified, graphed, and summarized. The Civil War furthered this orientation as doctors began to reify racial identity, thus negating human individuality. Doctors in the Confederate South did not record the names and identities of enslaved infants and children, treating them as property to use for harvesting vaccine matter.[8]

The rise of the International Sanitary Commission and the postwar efforts in the United States to thwart the 1866 cholera pandemic caused doctors to move aggressively toward analytical factors, from narrative mapping to statistical analysis, that further obscured the human beings whose bodies helped them pinpoint the presence of cholera.

Maladies of Empire shows how the use of numbers, which purported to be an objective, apolitical practice, became a central force in public health. Medical personnel counted the number of soldiers in a crowded cell, the number of enslaved Africans who died, the number of hospital workers who got sick, the number of people who died of yellow fever, the number of infected troops, the number of vaccinated enslaved children, the number of prisoners of war who died, the number of Muslim migrants in quarantine, in order to create a narrative about epidemics that otherwise created chaos and confusion. The numbers told a story that was thought to explain the cause and spread of an epidemic. *Maladies of Empire* shows how the methods of statistics, data collection, interviewing, and medical surveillance were spurred by imperialism, slavery, and war, all of which were based in violence.[9] Epidemiology resulted from larger aggressions against people and places that have been erased from its history.

Finally, I began this book long before the COVID-19 pandemic emerged in 2019. I hope that it helps readers understand that the epidemiological tools we depend on today originated among enslaved Africans on slave ships, colonized populations in the Caribbean and India, wartime casualties, prisoners of war, Muslim pilgrims, and other ordinary people. These populations were among the original case studies that enabled medical authorities to theorize about the cause, spread, and prevention of disease, but over time they have become ghosts disappearing into the darkness of the archives and replaced with theories and statistics. This book is an effort to turn the lights on and to see some of the people whose lives have led to the development of epidemiology, which continues to guide us through the pandemic today.

Notes

Introduction

1. This incident is recounted in Robert Dundas Thomson, "Clinical Observations," *Lancet* 32, no. 825 (June 22, 1839): 456–459. Direct quotations are from this article. Thomson, in turn, is reporting an account given by Thomas Trotter in his 1790 testimony before a committee of Parliament that was investigating the slave trade. The episode is also recounted in Markus Rediker, *The Slave Ship: A Human History* (New York: Penguin, 2007), 17–18. My telling imagines some details, to fill in the inevitable missing gaps.

2. See, for example, Charles Rosenberg, "The Therapeutic Revolution: Medicine, Meaning, and Social Change in Nineteenth-Century America," *Perspectives in Biology and Medicine* 20, no. 4 (1977): 485–506; John Harley Warner, *Therapeutic Perspective: Medical Practice, Knowledge, and Identity in America, 1820–1885* (Princeton, NJ: Princeton University Press, 1997).

3. See, for example, John Harley Warner, *Against the Spirit of System: The French Impulse in Nineteenth-Century American Medicine* (Princeton, NJ: Princeton University Press, 1998); Anne Hardy, "Cholera, Quarantine and the English Preventive System, 1850–1895," *Medical History* 37, no. 3 (1993): 250–269; Charles Rosenberg, *The Cholera Years: The United States in 1832, 1849, and 1866* (Chicago: University of Chicago Press, 1962); William Coleman, *Yellow Fever in the North: The Methods of Early Epidemiology* (Madison: University of Wisconsin Press, 1987). Some recent works that have investigated other sites of medical knowledge production include Rana Hogarth, *Medicalizing Blackness: Making Racial Difference in the Atlantic World* (Chapel Hill: University of North Carolina Press, 2017); Pablo Gómez, *The Experiential Caribbean: Creating Knowledge and Healing*

in the Early Modern Atlantic (Chapel Hill: University of North Carolina Press, 2017); Londa Schiebinger, *Secret Cures of Slaves: People, Plants, and Medicine in the Eighteenth-Century Atlantic World* (Stanford, CA: Stanford University Press, 2017).

4. See Ann Aschengrau and George R. Seage, *Essentials of Epidemiology in Public Health* (Burlington, MA: Jones and Bartlett Learning, 2018), 5–6.

5. Michel Foucault, *The History of Sexuality,* vol. 1: *An Introduction* (New York: Vintage, 1990), 140; Greta LaFleur and Kyla Schuller, eds., "Origins of Biopolitics in the Americas," special issue, *American Quarterly* 71, no. 3 (2019).

6. On the contribution of wartime medicine to medical knowledge, see Shauna Devine, *Learning from the Wounded: The Civil War and the Rise of American Medical Science* (Chapel Hill: University of North Carolina Press, 2014).

7. With the exception of a few recent studies, the importance of military medicine has been largely overlooked. For the British Empire, see Catherine Kelly, *War and the Militarization of British Army Medicine, 1793–1830* (London: Pickering and Chatto, 2011); Erica Charters, *Disease, War, and the Imperial State: The Welfare of the British Armed Forces during the Seven Years' War;* Mark Harrison, *Medicine in an Age of Commerce and Empire: Britain and Its Tropical Colonies, 1660–1830* (New York: Oxford University Press, 2010).

8. See the volumes of *Transactions of the Epidemiological Society of London* from the 1860s.

9. David Livingstone, *Putting Science in Its Place: Geographies of Scientific Knowledge* (Chicago: University of Chicago Press, 2003).

10. My analysis here draws on Charles Rosenberg and Janet Golden's formulation of "framing disease," which refers to the ways that technological and rhetorical frameworks—from the microscope to germ theory—make disease visible. Charles E. Rosenberg and Janet Golden, *Framing Disease: Studies in Cultural History* (New Brunswick, NJ: Rutgers University Press, 1997).

11. For some other works on science, medicine, and imperialism, see Richard H. Grove, *Green Imperialism: Colonial Expansion, Tropical Island Edens and the Origins of Environmentalism, 1600–1860* (Cambridge: Cambridge University Press, 1995); Londa Schiebinger and Claudia Swan, eds., *Colonial Botany: Science, Commerce, and Politics in the Early Modern World* (Philadelphia: University of Pennsylvania Press, 2007); J. R. McNeill, *Mosquito Empires: Ecology and War in the Greater Caribbean, 1620–1914* (Cambridge: Cambridge University Press, 2010); Mariola Espinosa, *Epidemic Invasions: Yellow Fever and the Limits of Cuban Independence, 1878–1930* (Chicago: University of Chicago Press, 2009); Hogarth, *Medicalizing Blackness;* Gómez, *Experiential Caribbean;* Juanita de Barros, *Reproducing the British Caribbean: Sex, Gender, and Population Politics after Slavery* (Chapel Hill: University of North Carolina Press, 2014).

12. On the literary and historical recovery of Black women, see Hazel V. Carby, *Reconstructing Womanhood: The Emergence of the Afro-American Woman Novelist* (New York: Oxford University Press, 1990); Evelyn Brooks Higginbotham, "African American Women's History and the Metalanguage of Race," *Signs* 17, no. 2 (1992): 251–274; Evelynn Hammonds, "Black (W)holes and the Geometry of Black Female Sexuality," *differences* 6, no. 2–3 (1994): 126–146; Valerie Smith, *Not Just Race, Not*

Just Gender: Black Feminist Readings (New York: Routledge, 1998); Farah Jasmine Griffin, *Beloved Sisters and Loving Friends: Letters from Rebecca Primus of Royal Oak, Maryland and Addie Brown of Hartford, Connecticut, 1854–1868* (New York: Ballantine Books, 2001). On more contemporary efforts to theorize about the relationship between subjectivity and the archive as well as critical fabulation, see Saidiya Hartman, "Venus in Two Acts," *Small Axe* 12, no. 2 (2008): 1–14; Marisa J. Fuentes, *Dispossessed Lives: Enslaved Women, Violence, and the Archive* (Philadelphia: University of Pennsylvania Press, 2016); Jennifer L. Morgan, "Accounting for 'The Most Excruciating Torment': Gender, Slavery, and Trans-Atlantic Passages," *History of the Present* 6, no. 2 (2016): 184–207.

1. Crowded Places

1. This account is told in [Robert John Thornton], *The Philosophy of Medicine, or Medical Extracts on the Nature of Health and Disease . . .* , 4th ed., vol. 1 (London: C. Whittingham, 1799), 328–330. Thornton's information came from John Z. Holwell's account of the incident in *A Genuine Narrative of the English Gentlemen, and Others, Who Were Suffocated in the Black-Hole in Fort-William, at Calcutta . . .* (London: Printed for A. Millar, 1758). Later historians have questioned the veracity of this account. See, for example, Partha Chatterjee, *The Black Hole of Empire* (Princeton, NJ: Princeton University Press, 2012), chap. 10.

2. [Thornton], *Philosophy of Medicine,* 1:325, italics in original.

3. For a discussion of medical understandings of air, see Mark Harrison, *Medicine in an Age of Commerce and Empire: Britain and Its Tropical Colonies* (Oxford: Oxford University Press, 2010), 32–45, 59–62.

4. John Pringle, *Observations on the Diseases of the Army in Camp and Garrison,* 6th ed. (London, 1766), 84, 110. For more on Pringle, see Harrison, *Medicine in an Age of Commerce and Empire,* 65–69.

5. [Thornton], *Philosophy of Medicine,* 1:325–330.

6. Ian Alexander Porter, "Thomas Trotter, M.D., Naval Physician," *Medical History* 7, no. 2 (1963): 154–164; Brian Vale and Griffith Edwards, *Physician to the Fleet: The Life and Times of Thomas Trotter, 1760–1832* (Woodbridge, Suffolk: Boydell Press, 2011).

7. On the history of the *Brookes* (or *Brooks,* as it is sometimes spelled), see Nicholas Radburn and David Eltis, "Visualizing the Middle Passage: The *Brooks* and the Reality of Ship Crowding in the Transatlantic Slave Trade," *Journal of Interdisciplinary History* 49, no. 4 (2019): 533–565; Stephen R. Berry, *A Path in the Mighty Waters: Shipboard Life and Atlantic Crossings to the New World* (New Haven, CT: Yale University Press, 2015), 28–31; Marcus Rediker, *The Slave Ship: A Human History* (New York: Penguin, 2008), chap. 10; Manisha Sinha, *The Slave's Cause: A History of Abolition* (New Haven: Yale University Press, 2016), 99–103.

8. Thomas Trotter, *Observations on the Scurvy: With a Review of the Opinions Lately Advanced on That Disease, and a New Theory Defended . . .* , 2nd ed. (London: Printed for T. Longman, 1792); [Thornton], *Philosophy of Medicine,* 1:331–333.

9. Trotter, *Observations on the Scurvy,* 51–55.

10. Trotter, *Observations on the Scurvy,* 52–53.

11. Trotter, *Observations on the Scurvy*, 53–55.
12. Trotter, *Observations on the Scurvy*, 55–56.
13. Trotter, *Observations on the Scurvy*, 57.
14. Trotter, *Observations on the Scurvy*, 60.
15. It is likely that there was a translator aboard. See W. O. Blake, *The History of Slavery and the Slave Trade* (Columbus, OH: J. & H. Miller, 1857), 127.
16. Trotter, *Observations on the Scurvy*, 59.
17. Trotter, *Observations on the Scurvy*, 69; see also Porter, "Thomas Trotter, M.D."
18. Trotter, *Observations on the Scurvy*, 137–138.
19. Trotter, *Observations on the Scurvy*, 224.
20. Trotter, *Observations on the Scurvy*, 69.
21. Trotter, *Observations on the Scurvy*, 69, 70 (quotation).
22. On the history of enslaved people as commodities and the health consequences of the international slave trade, see Sowande' M. Mustakeem, *Slavery at Sea: Terror, Sex, and Sickness in the Middle Passage* (Urbana: University of Illinois Press, 2016); Rediker, *The Slave Ship*.
23. Trotter, *Observations on the Scurvy*, xix.
24. Trotter, *Observations on the Scurvy*, 59.
25. Trotter, *Observations on the Scurvy*, 222–224. On Lind, see R. E. Hughes, "James Lind and the Cure of Scurvy," *Medical History* 19, no. 4 (1975): 342–351.
26. Trotter, *Observations on the Scurvy*, 62, 222–225.
27. Paul Farmer, "An Anthropology of Structural Violence," *Current Anthropology* 45, no. 3 (2004): 305–325.
28. Trotter, *Observations on the Scurvy*, 62–63.
29. Trotter, *Observations on the Scurvy*, 240–242.
30. For more on Trotter and scurvy, including his ideas relating to pneumatic chemistry, see Mark Harrison, "Scurvy on Sea and Land: Political Economy and Natural History, c. 1780–c. 1850," *Journal for Maritime Research* 15, no. 1 (2013): 7–25; Kenneth J. Carpenter, *Scurvy and Vitamin C* (Cambridge: Cambridge University Press, 1986), 88–90. Harrison does not mention that some of Trotter's observations were of enslaved Africans, but Carpenter does.
31. Trotter, *Observations on the Scurvy*, 240–241.
32. Trotter, *Observations on the Scurvy*, xxvii, 139–141, 68.
33. Trotter, *Observations on the Scurvy*, xix.
34. Historians have shown that eighteenth-century doctors connected crowded conditions in jails and hospitals to the outbreak of disease, but they have not paid attention to evidence from the international slave trade. See Erica Charters on British military physicians' contention that crowded conditions during the Seven Years' War led to typhus and dysentery: *Disease, War, and the Imperial State: The Welfare of the British Armed Forces during the Seven Years' War* (Chicago: University of Chicago Press, 1984), 87.
35. [Thornton], *Philosophy of Medicine*, 1:331, 334.
36. In the early nineteenth century physicians discussed the chemical nature of the air in crowded jails, hospitals, and ships. See, for example, Franklin Bache, *A System of*

Chemistry for the Use of Students of Medicine (Philadelphia: Printed and published for the author, 1819), 211. Popular discourse about the need for ventilation in homes does not seem to have developed until later in the nineteenth century. See Nancy Tomes, *Gospel of Germs: Men, Women, and the Microbe in American Life* (Cambridge: Harvard University Press, 1998); Alan M. Kraut, *Silent Travelers: Germs, Genes, and the Immigrant Menace* (Baltimore: Johns Hopkins University Press, 1995).

37. Quoted in [Thornton], *Philosophy of Medicine*, 1:333. See also William Bell Crafton, *A Short Sketch of the Evidence for the Abolition of the Slave Trade, Delivered before a Committee of the House of Commons . . .* (London, 1792), 8–9.

38. See Rediker, *The Slave Ship*, 332, on the importance of Trotter's testimony.

39. Trotter, *Observations on the Scurvy*, 243.

40. [Robert John Thornton], *Medical Extracts: On the Nature of Health, with Practical Observations: and The Laws of the Nervous and Fibrous Systems,* new edition, vol. 1 (London: J. Johnson, 1796), table of contents for part 2, section 2.

41. John Howard, *The State of the Prisons in England and Wales: With Preliminary Observations, and an Account of Some Foreign Prisons* (Warrington: William Eyres, 1777), 1; R. M. Gover, "Remarks on the History and Discipline of English Prisons in Some of Their Medical Aspects," *Lancet* 146, no. 3763 (October 12, 1895): 909–911.

42. Howard, *The State of the Prisons in England and Wales*, 12–13.

43. Howard, *The State of the Prisons in England and Wales*, 16–17.

44. Kevin Siena, *Rotten Bodies: Class and Contagion in Eighteenth-Century Britain* (New Haven, CT: Yale University Press, 2019).

45. Howard, *The State of the Prisons in England and Wales*, 19.

46. Howard, *The State of the Prisons in England and Wales*, 43.

47. John Howard, *The State of the Prisons in England and Wales: With Preliminary Observations, and an Account of Some Foreign Prisons and Hospitals*, 4th ed. (London: J. Johnson, C. Dilly, and T. Cadell, 1792), 471.

48. Simon Devereaux, "The Making of the Penitentiary Act, 1775–1779," *Historical Journal* 42, no. 2 (1999): 405–433; "John Howard and Prison Reform," Police, Prisons and Penal Reform, UK Parliament website, https://www.parliament.uk/about/living-heritage/transformingsociety/laworder/policeprisons/overview/prisonreform/.

49. Howard, *The State of the Prisons in England and Wales*, 4th ed., 93, 94; Isabel De Madariaga, *Politics and Culture in Eighteenth-Century Russia* (New York: Routledge, 1998), 119.

50. On how the debate over prisons offers an important site to chart the rise of medical ideas, see Siena, *Rotten Bodies*, 2–3.

51. For a discussion of the significance of past knowledge systems that are now regarded as pseudoscience, see Britt Rusert, *Fugitive Science: Empiricism and Freedom in Early African American Culture* (New York: New York University Press, 2017).

52. In the late nineteenth century, the medical community turned to evidence from the seventeenth, eighteenth, and early nineteenth centuries to illustrate how sanitation and ventilation prevented the spread of infectious disease and epidemics in jails. See, for example, H. D. Dudgeon, "Small-pox Manufactories in the Reign of

George III," *The Vaccination Inquirer and Health Review* 2, no. 16 (July 1880): 69. My analysis here is much indebted to Charles E. Rosenberg and Janet Golden's theory of "framing disease," which refers to the ways discursive, technological, and rhetorical frameworks redefine ideas about the pathology, etiology, and treatment of disease. Rosenberg and Golden, *Framing Disease: Studies in Cultural History* (New Brunswick, NJ: Rutgers University Press, 1997).

53. Gover, "Remarks on the History and Discipline of English Prisons."

54. Michael Worboys has questioned the concept of the bacterial revolution; see Worboys, "Was There a Bacteriological Revolution in Late Nineteenth-Century Medicine?" *Studies in History and Philosophy of Biological and Biomedical Sciences* 38, no. 1 (2007): 20–42.

55. Gover, "Remarks on the History and Discipline of English Prisons."

56. Edwin Chadwick, *Report . . . on an Inquiry into the Sanitary Conditions of the Labouring Population of Great Britain* (London, W. Clowes and Sons, 1842), 172–173.

57. John Howard, *An Account of the Principal Lazarettos in Europe . . . ,* 2nd ed., with additions (London: Printed for J. Johnson, D. Dilly, and T. Cadell, 1791), 53. Also see H. O. Lancaster, *Quantitative Methods in Biological and Medical Sciences: A Historical Essay* (New York: Springer-Verlag, 2012), 113.

58. Quoted in Edwin H. Ackerknecht, *Medicine at the Paris Hospital* (Baltimore: Johns Hopkins University Press, 1967), 16.

59. Howard, *The State of the Prisons in England and Wales*, 82.

60. Alain Corbin, *The Foul and the Fragrant: Odor and the French Social Imagination* (Cambridge, MA: Harvard University Press, 1986), 18, 24, 103, 102–105.

61. Ackerknecht, *Medicine at the Paris Hospital*, 61–62, 68.

62. "François Joseph Victor Broussais (1772–1838), System of Physiological Medicine," *Journal of the American Medical Association* 209, no. 10 (1969): 1523; Erwin H. Ackerknecht, "Broussais, or a Forgotten Medical Revolution," *Bulletin of the History of Medicine* 27, no. 4 (1953): 320–343.

63. F. J. V. Broussais, *Principles of Physiological Medicine . . . ,* trans. Isaac Hays and R. Eglesfeld Griffith (Philadelphia: Carey & Lea, 1832), 257.

64. Ackerknecht, *Medicine at the Paris Hospital*, 67.

65. Manfred J. Waserman and Virginia Kay Mayfield, "Nicolas Chervin's Yellow Fever Survey, 1820–1822," *Journal of the History of Medicine and Allied Sciences* 26, no. 1 (1971): 40–51; Mónica García, "Histories and Narratives of Yellow Fever in Latin America," in *The Routledge History of Disease,* ed. Mark Jackson (New York: Routledge, 2017), 232.

66. Ackerknecht, *Medicine at the Paris Hospital*, 158.

67. Ackerknecht, *Medicine at the Paris Hospital*, 122.

68. Ackerknecht, *Medicine at the Paris Hospital*, xi–xiii, 8–12, 117–134.

69. Bruno Latour, *The Pasteurization of France* (Cambridge: Harvard University Press, 1988), 22–25.

70. Colin Chisholm, *An Essay on the Malignant Pestilential Fever: Introduced into the West Indian Islands from Boullam, on the Coast of Guinea, as It Appeared in 1793, 1794, 1795, and 1796 . . . ,* 2nd ed., 2 vols. (London: Mawman, 1801), 2:2–215.

71. Katherine Arner, "Making Yellow Fever American: The Early American Republic, the British Empire and the Geopolitics of Disease in the Atlantic World," *Atlantic Studies* 7, no. 4 (2010): 449–450, 459–460; Jan Golinski, "Debating the Atmospheric Constitution: Yellow Fever and the American Climate," *Eighteenth-Century Studies* 49, no. 2 (2016): 156.

72. See Christopher Hamlin, "Predisposing Causes and Public Health in Early Nineteenth-Century Medical Thought," *Social History of Medicine* 5, no. 1 (1992): 43–70; Siena, *Rotten Bodies*.

73. Chisholm, *An Essay on the Malignant Pestilential Fever*, 2:9–10. Chisholm goes on to provide guidelines on establishing quarantines for ships that may carry disease or contain sick passengers. On stricken ships, sick passengers should be isolated from the rest of the crew, the interior of the ship scrubbed and sprinkled with vinegar, and the clothes and bedding of sick people burned (pp. 12–19). These provisions fall in line with arguments that Mark Harrison makes about how commerce and maritime culture served as a way for the medical community to consider the spread of infectious disease. See Harrison, *Medicine in an Age of Commerce*.

74. For British doctors' examinations of the environment, see Harrison, *Medicine in an Age of Commerce and Empire*.

75. David Arnold explains how much of the subcontinent remained outside the control of British medicine. Arnold, *Colonizing the Body: State Medicine and Epidemic Disease in Nineteenth-Century India* (Berkeley: University of California Press, 2002), 59, 98–115. On the use of statistical thinking as a form of colonization in India, see Gyan Prakash, *Another Reason: Science and the Imagination of Modern India* (Princeton, NJ: Princeton University Press, 1999).

76. D. Grierson, "Special Considerations on the Health of European Troops," *Transactions of the Medical and Physical Society of Bombay* 7 (1861): 1–44.

77. Grierson, "Special Considerations on the Health of European Troops," 22–24.

78. Grierson, "Special Considerations on the Health of European Troops," 22.

79. Grierson, "Special Considerations on the Health of European Troops," 28.

80. Grierson, "Special Considerations on the Health of European Troops," 31.

81. Hendrik Hartog, "Pigs and Positivism," *Wisconsin Law Review* 1985, no. 4 (1985): 899–935; Tomes, *Gospel of Germs*; Latour, *The Pasteurization of France*.

82. While some historians have conducted studies that revealed the practices made by the French hygienists as a necessary historical context for bacteriology, even this antecedent has a predecessor. See Latour, *The Pasteurization of France*.

83. The paper was read in 1741 to the Royal Society and published in 1743. Stephen Hales, *A Description of Ventilators: Whereby Great Quantities of Fresh Air May with Ease be Conveyed into Mines, Goals [sic], Hospitals, Work-houses and Ships, in Exchange for Their Noxious Air . . .* (London: Printed for W. Innys [etc.], 1743), 35.

84. Stephen Hales, *An Account of a Useful Discovery to Distill Double the Usual Quantity of Seawater . . . and An Account of the Great Benefit of Ventilators in Many Instances, in Preserving the Health and Lives of People, in Slave and Other Transport Ships, Which Were Read before the Royal Society . . .*, 2nd ed. (London: Printed for Richard Manby, 1756), 41.

85. Stephen Hales, *A Treatise on Ventilators: Wherein an Account Is Given of the Happy Effects of Several Trials That Have Been Made of Them. . . .* Part Second (London: Printed for Richard Manby, 1758), 3–4.

86. In the early twentieth century, some scholars traced the development of public health to Hales's invention of the ventilator. While I agree with their analysis, I highlight that Hales emphasized the international slave trade and enslaved Africans as crucial evidence to his arguments. See D. Fraser Harris, "Stephen Hales, the Pioneer in the Hygiene of Ventilation," *Scientific Monthly* 3, no. 5 (1916): 440–454.

87. Hales, *A Treatise on Ventilators,* 86.

88. Hales, *A Treatise on Ventilators,* 94.

89. Hales, *An Account of a Useful Discovery,* 43.

90. Stephanie E. Smallwood explains how the international slave trade tested the limits of human capacity in terms of the violence that enslaved Africans endured. Smallwood, *Saltwater Slavery: A Middle Passage from Africa to American Diaspora* (Cambridge: Harvard University Press, 2007).

91. Historians have identified a different relationship between slavery and prisons in a later period, arguing that slavery was a precursor to the rise of mass incarceration in the United States. See Talitha Leflouria, *Chained in Silence: Black Women and Convict Labor in the New South* (Chapel Hill: University of North Carolina Press, 2016); David M. Oshinsky, *Worse than Slavery: Parchman Farm and the Ordeal of Jim Crow Justice* (New York: Free Press, 1996).

92. Edwin Chadwick, *A Supplementary Report on the Results of a Special Inquiry into the Practice of Internment in Towns* (London: W. Clowes and Sons, 1843), 19.

2. Missing Persons

1. On quarantine in Malta, see Alexander Chase-Levenson, "Early Nineteenth-Century Mediterranean Quarantine as a European System," in *Quarantine: Local and Global Histories,* ed. Alison Bashford (London: Palgrave, 2016), 35–53; Alex Chase-Levenson, *The Yellow Flag: Quarantine and the British Mediterranean World, 1780–1860* (Cambridge: Cambridge University Press, 2020).

2. On washerwoman techniques, see Kathleen M. Brown, *Foul Bodies: Cleanliness in Early American Society* (New Haven, CT: Yale University Press, 2009), 30–32.

3. Arthur Todd Holroyd, *The Quarantine Laws, Their Abuses and Inconsistencies: A Letter Addressed to the Rt. Hon. Sir John Cam Hobhouse . . .* (London: Simpkin, Marshall & Co., 1839), 50. The passage on laundresses was quoted in a review of Holroyd's book in the *Lancet:* Review of *The Quarantine Laws,* by Arthur T. Holroyd, *Lancet* 31, no. 805 (Feb. 2, 1839): 702.

4. My analysis in this chapter draws on practitioners of Black feminist theory and criticism, who have recovered Black women's subjectivity from literature and the archives. See, for example, Hazel V. Carby, *Reconstructing Womanhood: The Emergence of the Afro-American Woman Novelist* (New York: Oxford University Press, 1987); Toni Morrison, *Playing in the Dark: Whiteness and the Literary Imagination* (Cambridge, MA: Harvard University Press, 1992); Valerie Smith,

Not Just Race, Not Just Gender: Black Feminist Readings (New York: Routledge, 1998); Farah Jasmine Griffin, ed., *Beloved Sisters and Loving Friends: Letters from Rebecca Primus of Royal Oak, Maryland and Addie Brown of Hartford, Connecticut, 1854–1868* (New York: Knopf, 1999); Saidiya Hartman, "Venus in Two Acts," *Small Axe* 12, no. 2 (2008): 1–14; Marisa J. Fuentes, *Dispossessed Lives: Enslaved Women, Violence, and the Archive* (Philadelphia: University of Pennsylvania Press, 2016).

5. For a comprehensive study of European responses to quarantine, see Peter Baldwin, *Contagion and the State in Europe, 1830–1930* (Cambridge: Cambridge University Press, 1999). There is a growing literature on quarantine outside the metropole; see, for example, Alison Bashford, *Imperial Hygiene: A Critical History of Colonialism, Nationalism and Public Health* (New York: Palgrave Macmillan, 2004); John Chircop and Francisco Javier Martinez, eds., *Mediterranean Quarantines, 1750–1914* (Manchester: Manchester University Press, 2018).

6. See, for example, A[mariah] Brigham, *Treatise on Epidemic Cholera; Including an Historical Account of Its Origin and Progress, . . .* (Hartford: H. and F. J. Huntington, 1832), 350; Gavin Milroy, *Quarantine and the Plague: Being a Summary of the Report on These Subjects Recently Addressed to the Royal Academy of Medicine in France; with Introductory Observations, Extracts from Parliamentary Correspondence, and Notes* (London: Samuel Highley, 1846); Gavin Milroy, *Quarantine as It Is, and as It Ought to Be* (London: Savill & Edwards, 1859). On the history of debates over quarantine, see Mark Harrison, *Contagion: How Commerce Spread Disease* (New Haven, CT: Yale University Press, 2012).

7. Historians have pointed out that medical professionals were not neatly divided as either contagionists or anticontagionists but were, in fact, much more nuanced in their thinking about the cause of epidemics. See, for example, Margaret Pelling, *Cholera, Fever and English Medicine* (New York: Oxford University Press, 1978); Christopher Hamlin, "Predisposing Causes and Public Health in Early Nineteenth-Century Medical Thought," *Social History of Medicine* 5, no. 1 (1992), 43–70.

8. G. C. Cook, "William Twining (1790–1835): The First Accurate Clinical Description of 'Tropical Sprue' and Kala-Azar?" *Journal of Medical Biography* 9, no. 3 (August 2001): 125–131.

9. William Twining, *Clinical Illustrations of the More Important Diseases of Bengal, with the Result of an Inquiry into Their Pathology and Treatment* (Calcutta: Baptist Mission Press, 1832). For more on the history of cholera, see Christopher Hamlin, *Cholera: The Biography* (New York: Oxford University Press, 2009).

10. Twining, *Clinical Illustrations of the More Important Diseases of Bengal*, 535–536.

11. Twining, *Clinical Illustrations of the More Important Diseases of Bengal*, 536–538.

12. For an overview of anticontagionism, see Erwin H. Ackerknecht, "Anticontagionism between 1821 and 1867: The Fielding H. Garrison Lecture," *International Journal of Epidemiology* 38, no. 1 (2009): 7–21; Christopher Hamlin, "Commentary: Ackerknecht and 'Anticontagionism': A Tale of Two Dichotomies," *International Journal of Epidemiology* 38, no. 1 (2009): 22–27.

13. Brigham, *A Treatise on Epidemic Cholera*, 295–331.

14. Brigham, *A Treatise on Epidemic Cholera*, 301, 306, 322–328; quotes from 324, 323.

15. Brigham, *A Treatise on Epidemic Cholera*, 36; Charles Telfair, "Account of the Epidemic Cholera, as It Occurred at Mauritius: Communicated to Dr Macdonnel, Belfast," *Edinburgh Medical and Surgical Journal* 17, no. 69 (1821): 517–518. On the decline of the enslaved population on Mauritius due to the 1819 cholera epidemic, see Sadasivam Jaganada Reddi and Sheetal Sheena Sookrajowa, "Slavery, Health, and Epidemics in Mauritius 1721–1860," in *The Palgrave Handbook of Ethnicity*, ed. Steven Ratuva (Singapore: Springer Nature Singapore, 2019), 1749–1765.

16. Brigham, *A Treatise on Epidemic Cholera*, 331.

17. Brigham, *A Treatise on Epidemic Cholera*, 317–320.

18. Holroyd, *The Quarantine Laws*, 16, 18–19, 36–37.

19. Holroyd, *The Quarantine Laws*, 22. Henry Abbott was also a collector of Egyptian artifacts, which he displayed in New York City; see "Egypt on Broadway," New-York Historical Society blog post, http://blog.nyhistory.org/egypt-on-broadway/.

20. Holroyd, *The Quarantine Laws*, 25.

21. Samuel Henry Dickson, *Essays on Pathology and Therapeutics*, 2 vols. (Charleston: McCarter & Allen, 1845), 2:618–619.

22. This is not to argue that Dickson does not make claims about racial difference. He claims, for example, that the treatment for smallpox should include the warm bath as "one of the earliest measures in the management of negroes and of whites of the lower classes" and that the "rose colored spots" of typhus and typhoid are not seen in "negroes"; Dickson, *Essays on Pathology and Therapeutics*, 2:533, 547. Discussing the different races in general, he claims that Blacks and whites have different susceptibility to certain diseases, but the reason is unclear (vol. 1:26–27).

23. Dickson, *Essays on Pathology and Therapeutics*, 2:607–608.

24. M. L. Knapp, *Researches on Primary Pathology, and the Origin and Laws of Epidemics*, 2 vols. (Philadelphia: The Author, 1858), 1:229. Knapp was admitted to the Maryland Medical and Chirurgical Faculty on June 1, 1829. *Maryland Medical Recorder* 1, no. 1 (1829), 769.

25. See, for example, Rana A. Hogarth, *Medicalizing Blackness: Making Racial Difference in the Atlantic World, 1780–1840* (Chapel Hill: University of North Carolina Press, 2017).

26. On the rise of slavery in New Orleans and the domestic slave trade, see Walter Johnson, *Soul by Soul: Life inside the Antebellum Slave Market* (Cambridge, MA: Harvard University Press, 2000). On theories about the environment as a cause for epidemic disease, see Mark Harrison, *Medicine in an Age of Commerce and Empire: Britain and Its Tropical Colonies 1660–1830* (New York: Oxford University Press, 2010).

27. Holroyd, *The Quarantine Laws*, 16.

28. On medicine's shift from an empirical to a more scientific discipline, see W. F. Bynum, *Science and the Practice of Medicine in the Nineteenth Century* (Cambridge: Cambridge University Press, 1994); Mark Weatherall, "Making Medicine Scientific: Empiricism, Rationality, and Quackery in Mid-Victorian Britain," *Social History of Medicine* 9, no. 2 (1996): 175–194; Harold J. Cook, "The History of Medicine and the Scientific Revolution" *Isis* 102, no. 1 (2011): 102–108.

29. Holroyd, *The Quarantine Laws*, 17–18.
30. Holroyd, *The Quarantine Laws*, 18, 23, 27, 36.
31. Holroyd, *The Quarantine Laws*, 18.
32. The Arab villages represent an example of what theorist Edward Said refers to as "Orientalism," which means the systematic and often patronizing language Europeans used to describe the Middle East. It is why the image of the hut would have been familiar to many readers. Edward Said, *Orientalism* (New York: Vintage, 1979).
33. Holroyd, *The Quarantine Laws*, 31.
34. Holroyd, *The Quarantine Laws*, 27.
35. Holroyd, *The Quarantine Laws*, 29.
36. Harrison, *Medicine in an Age of Commerce and Empire*, 71, 75–76. British novels based on stories of adventure throughout the British Empire, from *Robinson Crusoe*, published in 1719, to *Treasure Island*, published in 1882, were runaway bestsellers. Military physicians may have understood their time abroad in this context. British colonial officials also operated as adventurous investigators in collecting and classifying plants, animals, and other natural history objects. See Miranda Carter, "British Readers and Writers Need to Embrace Their Colonial Past," *Guardian*, January 23, 2014; Mary Louise Pratt, *Imperial Eyes: Travel Writing and Transculturalism* (New York: Routledge, 2007).
37. Harrison, *Medicine in an Age of Commerce and Empire*, chap. 3.
38. Holroyd, *The Quarantine Laws*, 52–53.
39. Holroyd, *The Quarantine Laws*, 53–54.
40. Holroyd, *The Quarantine Laws*, 41–42.
41. Holroyd, *The Quarantine Laws*, 41, 43, 51. See also Chase-Levenson, *The Yellow Flag*, 99–104.
42. John Slight, *The British Empire and the Hajj: 1865–1956* (Cambridge, MA: Harvard University Press, 2015); Eileen Kane, *Russian Hajj: Empire and the Pilgrimage to Mecca* (Ithaca, NY: Cornell University Press, 2015).
43. Holroyd, *The Quarantine Laws*, 14, 29.
44. Holroyd, *The Quarantine Laws*, 39.
45. Milroy, *Quarantine and the Plague*.
46. Milroy, *Quarantine and the Plague*, 5.
47. Milroy, *Quarantine and the Plague*, 40–41.
48. Milroy, *Quarantine and the Plague*, 32. Birsen Bulmuş claims that Milroy and the French physicians who wrote the report on quarantine did use ethnic distinctions to explain different rates of mortality from plague, but as I show, their explanation focused on sanitary factors, not assumptions about racial inferiority. Bulmuş also claims that Milroy and other European doctors used their criticism of poor sanitary conditions to justify colonial rule, but these doctors and their political contemporaries did not need a medical argument to justify colonialism—they had brute force, political will, and elitism to fuel their campaigns. Alison Bashford offers a convincing way to interpret the relationship between the colonial doctors and subjugated populations by examining how many cases unfolding between 1850 and 1950 can be framed as examples of separating the pure from the infected, particularly in relation

to contagion and colonialism. See Bulmuş, *Plague, Quarantines and Geopolitics in the Ottoman Empire* (Edinburgh: Edinburgh University Press, 2012), 131–140; Bashford, *Imperial Hygiene*.

49. Milroy, *Quarantine and the Plague,* 6–9.
50. Milroy, *Quarantine and the Plague*, 8.

3. Epidemiology's Voice

1. See Sharla M. Fett, *Recaptured Africans: Surviving Slave Ships, Detention, and Dislocation in the Final Years of the Slave Trade* (Chapel Hill: University of North Carolina Press, 2017); Matthew S. Hopper, *Slaves of One Master: Globalization and Slavery in Arabia in the Age of Empire* (New Haven: Yale University Press, 2015).
2. [James Ormiston] McWilliam, *Report of the Fever at Boa Vista*, Presented to the House of Commons, in Pursuance of Their Address of 16th March, 1857 (London: Printed by T. R. Harrison), 76, 95, 110.
3. McWilliam, *Report of the Fever at Boa Vista*, 94, 109.
4. For accounts of the *Eclair*'s voyage see McWilliam, *Report of the Fever at Boa Vista,* 77–82; "Correspondence Respecting the History of the 'Eclair' Fever," *Medico-Chirurgical Review* 49 (July 1846): 235–246; Reviews, *Lancet* 50, no. 1255 (1847): 307–311; *British and Foreign Medico-Chirurgical Review,* 1, Art. 3 (January 1848): 49–79; Mark Harrison, *Contagion: How Commerce Spread Disease* (New Haven, CT: Yale University Press, 2012), 80–84; Lisa Rosner, "Policing Boundaries: Quarantine and Professional Identity in Mid Nineteenth-Century Britain," in *Mediterranean Quarantines, 1750–1914: Space, Identity and Power,* ed. John Chircop and Francisco Javier Martinez (Manchester: Manchester University Press, 2018), 125–144.
5. Harrison, *Contagion,* 82–84.
6. Harrison, *Contagion,* 94–97.
7. McWilliam, *Report of the Fever at Boa Vista,* 8, 94.
8. On the problem of uncovering archival evidence of enslaved people of African descent as patients, see Jim Downs, "#BlackLivesMatter: Toward an Algorithm of Black Suffering during the Civil War and Reconstruction," *J19: The Journal of Nineteenth-Century Americanists* 4, no. 1 (2016): 198–206.
9. James Ormiston McWilliam, *Medical History of the Expedition to the Niger during the Years 1841–2: Comprising an Account of the Fever Which Led to Its Abrupt Termination* (London: J. Churchill, 1843); review of *Medical History of the Expedition to the Niger,* by James Ormiston McWilliam, *Medico-Chirurgical Review* 39, no. 78 (October 1843): 377–384. For a biography of McWilliam, see R. R. Willcox, "James Ormiston McWilliam (1807–1862)," *Transactions of the Royal Society of Tropical Medicine and Hygiene* 44, no. 1 (1950): 127–144.
10. McWilliam, *Medical History of the Expedition to the Niger,* 27–29, 37.
11. McWilliam, *Medical History of the Expedition to the Niger,* frontispiece, 60.
12. McWilliam, *Medical History of the Expedition to the Niger,* 63, 254.
13. On the American medical profession and public's understanding of disease transmission in the first half of the nineteenth century, see Charles Rosenberg, *The*

Cholera Years: The United States in 1832, 1849, and 1866 (Chicago: University of Chicago Press, 1962).

14. Londa Schiebinger, *Plants and Empire: Colonial Bioprospecting in the Atlantic World* (Cambridge, MA: Harvard University Press, 2004). Feminist historians have argued that there is no clear division between nature and culture. Donna Haraway coined the term *natureculture* to describe their inseparability. Banu Subramaniam has advanced Haraway's analysis by calling for scholars to consider the biographies of scientists in order to uncover how their biases shape scientific knowledge production. McWilliam observes the interaction between nature and culture, and his documentation of his observations unwittingly illustrates Subramaniam's thesis by mixing his theories about the spread of fever with his other observations. Donna J. Haraway, *The Companion Species Manifesto: Dogs, People, and Significant Otherness* (Chicago: Prickly Paradigm Press, 2015); Banu Subramanian, *Ghost Stories for Darwin: The Science of Variation and the Politics of Diversity* (Urbana: University of Illinois Press, 2014).

15. McWilliam, *Medical History of the Expedition to the Niger,* 180.

16. McWilliam, *Medical History of the Expedition to the Niger,* 205–207.

17. McWilliam, *Medical History of the Expedition to the Niger,* 131–148, 180–181, 194–202; quote on 200.

18. McWilliam, *Medical History of the Expedition to the Niger,* 156.

19. McWilliam, *Medical History of the Expedition to the Niger,* 157, 159.

20. McWilliam, *Medical History of the Expedition to the Niger,* 161–175.

21. McWilliam, *Medical History of the Expedition to the Niger,* 162.

22. *Edinburgh New Philosophical Journal* 31 (April–October 1841): 183–184.

23. McWilliam, *Medical History of the Expedition to the Niger,* 171, 179, 180.

24. W. Burnett, "Instructions to Dr. McWilliam," in McWilliam, *Report of the Fever at Boa Vista,* 4–5; World Health Organization, "Contact Tracing," Newsroom, May 9, 2017, https://www.who.int/news-room/q-a-detail/contact-tracing.

25. For patients examined, McWilliam, *Report of the Fever at Boa Vista,* 94.

26. McWilliam, *Report of the Fever at Boa Vista,* 14.

27. McWilliam, *Report of the Fever at Boa Vista,* 14, 16.

28. McWilliam, *Report of the Fever at Boa Vista,* 16–17.

29. McWilliam, *Report of the Fever at Boa Vista,* 17–18.

30. McWilliam, *Report of the Fever at Boa Vista,* 18.

31. McWilliam, *Report of the Fever at Boa Vista,* 21–22.

32. McWilliam, *Report of the Fever at Boa Vista,* 22–24.

33. McWilliam, *Report of the Fever at Boa Vista,* 84.

34. McWilliam, *Report of the Fever at Boa Vista,* 23–24.

35. McWilliam, *Report of the Fever at Boa Vista,* 26–29.

36. See, for example, McWilliam, *Report of the Fever at Boa Vista,* 45, 49.

37. McWilliam, *Report of the Fever at Boa Vista,* 14, 16, 17.

38. McWilliam, *Report of the Fever at Boa Vista,* 82, 108.

39. On Portajo, McWilliam, *Report of the Fever at Boa Vista,* 23, 24, 32–33, 59, 60.

40. On Boaventura and Cabeçada, McWilliam, *Report of the Fever at Boa Vista,* 88–89.

41. McWilliam, *Report of the Fever at Boa Vista,* 32–33.

42. On Rosa Fortes, McWilliam, *Report of the Fever at Boa Vista*, 33–34, 47, 52, 55.

43. McWilliam, *Report of the Fever at Boa Vista*, 33–34.

44. For the testimony of those identified as slaves, McWilliam, *Report of the Fever at Boa Vista*, 16, 24, 32, 39, 54, 58–62.

45. McWilliam, *Report of the Fever at Boa Vista*, 104–105.

46. McWilliam, *Report of the Fever at Boa Vista*, 108, 111.

47. McWilliam, *Report of the Fever at Boa Vista*, 79, 104–105.

48. McWilliam, *Report of the Fever at Boa Vista*, 109–110, 111.

49. McWilliam, *Report of the Fever at Boa Vista*, 38.

50. McWilliam, *Report of the Fever at Boa Vista*, 112.

51. McWilliam listed many of the reviews in J. O. M'William, *Further Observations on That Portion of the Second Report on Quarantine by the General Board of Health, Which Relates to the Yellow Fever Epidemy on Board H.M.S. Eclair, and at Boa Vista in the Cape de Verde Islands* (London: William Tyler, 1852), 2.

52. Harrison, *Contagion*, 85–86, 97–98.

53. Harrison, *Contagion*, 98–99.

54. Gilbert King, *The Fever at Boa Vista in 1845–6, Unconnected with the Visit of the "Eclair" to That Island* (London: John Churchill, 1852).

55. Reviews, *Lancet* 50, no. 1255 (1847): 310.

56. McWilliam's report was widely reviewed and praised for its thoroughness; see, for example, *Medico-Chirurgical Review and Journal of Practical Medicine* 51 (July 1, 1847): 217–233. On how subjugated and enslaved populations have contributed to medical knowledge production, see Sharla Fett, *Working Cures: Healing, Health, and Power on Southern Slave Plantations* (Chapel Hill: University of North Carolina Press, 2002); Londa Schiebinger, *Secret Cures of Slaves: People, Plants, and Medicine in the Eighteenth-Century Atlantic World* (Stanford, CA: Stanford University Press, 2017); Pablo Gómez, *The Experiential Caribbean: Creating Knowledge and Healing in the Early Modern Atlantic* (Chapel Hill: University of North Carolina Press, 2017).

57. On emotional labor, see Arlie Russell Hochschild, *The Second Shift: Working Parents and the Revolution at Home* (London: Piatkus, 1990); Mary E. Guy, Meredith A. Newman, and Sharon H. Mastracci, *Emotional Labor: Putting the Service in Public Service* (Armonk, NY: M. E. Sharpe, 2008).

58. McWilliam, *Report of the Fever at Boa Vista*, 42–43.

59. Studies on the history of global health, and about medicine and imperialism, tend to begin in the late nineteenth century. See, for example, Randall Packard, *A History of Global Health: Interventions into the Lives of Other Peoples* (Baltimore: Johns Hopkins University Press, 2016); Warwick Anderson, *Colonial Pathologies: American Tropical Medicine, Race, and Hygiene in the Philippines* (Durham, NC: Duke University Press, 2006); David Arnold, *Colonizing the Body: State Medicine and Epidemic Disease in Nineteenth Century India* (Berkeley: University of California Press, 1993).

60. On Snow's methods, see Tom Koch, "John Snow, Hero of Cholera: RIP," *Canadian Medical Association Journal* 178, no. 13 (2008): 1736.

4. Recordkeeping

1. George McCall Theal, *History of South Africa since September 1795,* vol. 3 (London: S. Sonnenschein, 1908), 70–79; Hilary M. Cary, *Empire of Hell: Religion and the Campaign to End Convict Transportation in the British Empire, 1788–1875* (Cambridge: Cambridge University Press, 2019), 232–239.

2. On the hulks, see "Convict Hulks," Sydney Living Museum, https://sydneyliving museums.com.au/stories/convict-hulks.

3. *The Albion, A Journal of News, Politics, and Literature* (New York), November 24, 1849, 558; *Bombay Times and Journal of Commerce,* December 12, 1849, 860; *Maine Farmer,* February 21, 1850, 18, 8; *The Independent,* January 24, 1850, 2, 60; *The Spectator,* April 6, 1850, 319.

4. *Bombay Times and Journal of Commerce,* December 12, 1849, 860.

5. For an example of the resistance story, see South African History Online: Towards a People's History, https://www.sahistory.org.za/dated-event/neptune-288-convicts -board-enters-simons-bay-amid-strong-resistance-cape-inhabitants.

6. George B. Grundy to Sir George Grey, May 15, 1849, pp. 101–102, Letters from the Home Office and Treasury on Matters Relating to Bermuda, CO 37 / 130, National Archives, Kew. In 1842, when George Baxter Grundy was nineteen, he was charged with forgery: *Manchester Courier and Lancashire General Advertiser,* August 6, 1842; *Bolton Chronicle Greater Manchester,* August 6, 1842.

7. There is a long tradition among men to refer to their intimate relationships in terms of marriage. On the narrative contrivance of same-sex marriage in the first half of the nineteenth century, see Timothy Stewart-Winter and Simon Stern, "Picturing Same-Sex Marriage in the Antebellum United States: The Union of 'Two Most Excellent Men' in Longstreet's 'A Sage Conversation,'" *Journal of the History of Sexuality* 19, no. 2 (2010): 197–222.

8. For more on same-sex relationships between men on the ship, see Jim Downs, "The Gay Marriages of a Nineteenth-Century Prison Ship," *New Yorker,* July 2, 2020, https://www.newyorker.com/culture/culture-desk/the-gay-marriages-of-a-nineteenth -century-prison-ship.

9. Philip Harling, "The Trouble with Convicts: From Transportation to Penal Servitude, 1840–67," *Journal of British Studies* 53, no. 1 (2014): 80–110. On the marking of same-sex desire among men in the nineteenth century, see Jim Downs, "With Only a Trace: Same-Sex Sexual Desire and Violence on Slave Plantations, 1607–1865," in *Connexions: Histories of Race and Sex in North America,* ed. Jennifer Brier, Jim Downs, and Jennifer Morgan (Champaign: University of Illinois Press, 2016), 15–37.

10. *An Earnest and Respectful Appeal to the British and Foreign Bible Society, by Its South African Auxiliary, on Behalf of the Injured Colony of the Cape of Good Hope (with Reference to Convict Transportation)* (Cape Town: Saul Solomon & Co., 1849), 22–23.

11. Mark Harrison, *Medicine in an Age of Commerce and Empire: Britain and Its Tropical Colonies, 1660–1830* (New York: Oxford University Press, 2010), 44–45.

12. Oz Frankel, *States of Inquiry: Social Investigations and Print Culture in Nineteenth-Century Britain and the United States* (Baltimore: Johns Hopkins University Press, 2006).

13. E. Ashworth Underwood, "The History of Cholera in Great Britain," *Proceedings of the Royal Society of Medicine* 41, no. 3 (1948): 165–173; Charles Rosenberg, *The Cholera Years: The United States in 1832, 1849, and 1866* (Chicago: University of Chicago Press, 1962); David Arnold, "Cholera and Colonialism in British India," *Past & Present* no. 113 (1986): 118–151; Christopher Hamlin, *Cholera: The Biography* (New York: Oxford University Press, 2009).

14. For examples of physicians who followed this trajectory, see Harrison, *Medicine in an Age of Commerce and Empire.*

15. [Gavin Milroy], *Report on the Cholera in Jamaica, and on the General Sanitary Condition and Wants of the Island* (London: Eyre and Spottiswoode, 1853), 3.

16. On the global reaction to cholera, see Hamlin, *Cholera*, 4–21.

17. James Henry, August 4 1847–September 30, 1848, "Medical and Surgical Journal of Her Majesty's Sloop Antelope," ADM 101/85/3/4, National Archives, Kew. All quotations in this section are from Henry's journal; phrases rendered in italics are underlined in the original document. For an account by an assistant naval surgeon about cholera on an 1849 voyage to Hong Kong, see Bronwen E. J. Goodyer, "An Assistant Ship Surgeon's Account of Cholera at Sea," *Journal of Public Health* 30, no. 3 (2008): 332–338.

18. Christopher Hamlin discusses how the term "cholera" changed over time and how the disease was differentiated from other intestinal disorders in *Cholera*, 21–28.

19. Mark Harrison, "Science and the British Empire," *Isis* 96, no. 1 (2005): 56–63; Juanita de Barros, *Reproducing the British Caribbean: Sex, Gender, and Population Politics after Slavery* (Chapel Hill: University of North Carolina Press, 2014), 36.

20. For an analysis of the creation of scientific knowledge in the Caribbean during the long seventeenth century, see Pablo F. Gómez, *The Experiential Caribbean: Creating Knowledge and Healing in the Early Modern Atlantic* (Chapel Hill: University of North Carolina Press, 2017). Mark Harrison discusses British physicians' work in the Caribbean in *Medicine in an Age of Commerce and Empire.*

21. [Milroy], *Report on the Cholera in Jamaica*, 5–7. See also Deborah Jenson, Victoria Szabo, and the Duke FHI Haiti Humanities Laboratory Student Research Team, "Cholera in Haiti and Other Caribbean Regions, 19th Century," *Emerging Infectious Diseases* 17, no. 11 (2011): 2130–2135.

22. [Milroy], *Report on the Cholera in Jamaica*, 5.

23. *Report on the Epidemic Cholera of 1848 & 1849,* Presented to Both Houses of Parliament by Command of Her Majesty (London: W. Clowes & Sons, 1850). For other secondary sources on Milroy's report about Jamaica, see C. H. Senior, "Asiatic Cholera in Jamaica (1850–1855)," *Jamaica Journal* 26, no. 2 (December 1997): 25–42; Christienna D. Fryar, "The Moral Politics of Cholera in Postemancipation Jamaica," *Slavery and Abolition* 34, no. 4 (2013): 598–618; de Barros, *Reproducing the British Caribbean*; Aaron Graham, "Politics, Persuasion and Public Health in Jamaica, 1800–1850," *History* 104, no. 359 (2019): 63–82.

24. Rita Pemberton, "Dirt, Disease and Death: Control, Resistance and Change in the Post-Emancipation Caribbean," *História, Ciências, Saúde-Manguinhos* 19, suppl. 1 (2012): 47–58.

25. Gavin Milroy to Benjamin Hawes, February 10, 1851, Lucea, Jamaica, CO 318 / 194, National Archives, Kew.

26. See, for example, Rosenberg, *The Cholera Years,* on the epidemic in New York City. As we have seen in previous chapters, however, physicians did frequently learn from their colleagues, as in the case of Edwin Chadwick, who collected data about the sanitary conditions of towns in England.

27. [Milroy], *Report on the Cholera in Jamaica,* 3–4.

28. [Milroy], *Report on the Cholera in Jamaica,* 42–43.

29. [Milroy], *Report on the Cholera in Jamaica,* 42–43.

30. [Milroy], *Report on the Cholera in Jamaica,* 42–45.

31. [Milroy], *Report on the Cholera in Jamaica,* 10–11.

32. [James] Watson, "Cholera in Jamaica. An Account of the First Outbreak of the Disease in That Island in 1850," *Lancet* 57, no. 1428 (1851): 40–41.

33. Watson, "Cholera in Jamaica."

34. Watson, "Cholera in Jamaica."

35. [Milroy], *Report on the Cholera in Jamaica,* 38.

36. [Milroy], *Report on the Cholera in Jamaica,* 14, 16.

37. Edwin Chadwick, *Report . . . on an Inquiry into the Sanitary Conditions of the Labouring Population of Great Britain* (London, W. Clowes and Sons, 1842). On Chadwick's influence on epidemiology, see *Companion Encyclopedia of the History of Medicine,* vol. 2, ed. W. F. Bynum and Roy Porter (London: Routledge, 1993), 1242–1244.

38. Gavin Milroy to Edwin Chadwick, March 10, 1851, CO 318 / 194, National Archives, Kew.

39. This also undermines the false dichotomy that many scholars imply about how British physicians viewed "the Tropics" as starkly different from the metropole. Milroy believed that the principles of sanitation transcended place. He believed that Chadwick's recommendations, which grew out of problems in England, could be applied to the Caribbean. On the persistence of the "Tropics" in the European imagination, see Nancy Leys Stepan, *Picturing Tropical Nature* (Ithaca, NY: Cornell University Press, 2001).

40. Margaret Jones, *Public Health in Jamaica, 1850–1940: Neglect, Philanthropy and Development* (Kingston: University of West Indies Press, 2013); Fryar, "The Moral Politics of Cholera." For more general works on how structural violence invariably produces the conditions that facilitate the spread of epidemics, see Johan Galtung, "Violence, Peace, and Peace Research," *Journal of Peace Research* 6, no. 3 (1969): 167–191; Paul Farmer, "An Anthropology of Structural Violence," *Current Anthropology* 45, no. 3 (2004), 305–325.

41. Milroy to Chadwick, March 10, 1851.

42. [Milroy], *Report on the Cholera in Jamaica,* 108.

43. [Milroy], *Report on the Cholera in Jamaica,* 112 (appendix A).

44. [Milroy], *Report on the Cholera in Jamaica,* 110–115 (appendix A).
45. Scholars have recently pointed out how colonialism and imperialism produced disruptions in the environment that caused disease outbreaks in the Caribbean. On the adverse health and environmental consequences of colonialism and imperialism, see, for example, Mariola Espinosa, *Epidemic Invasions: Yellow Fever and the Limits of Cuban Independence* (Chicago: University of Chicago Press, 2009); J. R. McNeill, *Mosquito Empires: Ecology and War in the Greater Caribbean, 1620–1914* (Cambridge: Cambridge University Press, 2010); Richard Grove, *Green Imperialism: Colonial Expansion, Tropical Island Edens and the Origins of Environmentalism, 1600–1860* (Cambridge: Cambridge University Press, 1996).
46. [Milroy], *Report on the Cholera in Jamaica,* 111 (appendix A).
47. [Milroy], *Report on the Cholera in Jamaica,* 99–105.
48. Milroy to Chadwick, March 10, 1851.
49. [Milroy], *Report on the Cholera in Jamaica,* 42.
50. Thomas C. Holt, *The Problem of Freedom: Race, Labor, and Politics in Jamaica and Britain, 1832–1938* (Baltimore: Johns Hopkins University Press, 1992), 56, 133–167.
51. Pemberton, "Dirt, Disease and Death."
52. Gavin Milroy to Benjamin Hawes, May 28, 1851, Kingston, Jamaica, CO 318 / 194, National Archives, Kew.
53. On the public health aftermath of the epidemic, see Fryar, "The Moral Politics of Cholera in Postemancipation Jamaica"; Graham, "Politics, Persuasion and Public Health in Jamaica, 1800–1850"; Senior, "Asiatic Cholera in Jamaica." Fryar argues that Milroy's sanitary campaign operated under a moral impulse. She argues that Milroy used the cholera epidemic to indict Black people and then called for their sanitary improvement in an effort to reform their moral conduct. However, Milroy's comments must be placed within the broader global context of sanitary reform. British medical professionals referred to many places as filthy and unsanitary; this was not limited to Jamaica. While moral judgments shaped their indictments, Milroy and other physicians were more invested in understanding how disease spreads than in launching a moral crusade.
54. "Medical News: The Epidemiological Society, Dec. 4, 1851," *Lancet* 58, no. 1476 (December 13, 1851), 568.
55. On Milroy's career, see "Obituary—Gavin Milroy," *British Medical Journal* 1, no. 1313 (1886): 425–426; Mark Harrison, "Gavin Milroy (1805–1886)," *Oxford Dictionary of National Biography Online,* updated September 23, 2010.
56. Gavin Milroy, "Address at the Opening of the Session, 1864–65," *Transactions of the Epidemiological Society of London,* vol. 2 (London: Robert Hardwicke, 1867), 247–256.
57. [Milroy], *Report on the Cholera in Jamaica,* 115 (appendix A).
58. My idea here develops from historian Ann Blair's analysis of knowledge production in the early modern period; Ann M. Blair, *Too Much to Know: Managing Scholarly Information before the Modern Age* (New Haven, CT: Yale University Press, 2011).

59. In *States of Inquiry*, Oz Frankel has analyzed the role of the state as a key player in publishing social investigations that shaped print culture in both Britain and the United States. Frankel emphasizes the production of reports on British domestic issues, like child labor and poverty, while my focus is on imperialism and war.

5. Florence Nightingale

1. On the American Civil War as a biological war, see Jim Downs, *Sick from Freedom: African American Illness and Suffering during the Civil War and Reconstruction* (New York: Oxford University Press, 2012); Shauna Devine, *Learning from the Wounded: The Civil War and the Rise of American Medical Science* (Chapel Hill: University of North Carolina Press, 2014). On disease and the Cuban insurrection, see Matthew Smallman-Raynor and Andrew D. Cliff, "The Spatial Dynamics of Epidemic Diseases in War and Peace: Cuba and the Insurrection against Spain, 1895–98," *Transactions of the Institute of British Geographers* 24, no. 3 (1999): 331–350. On the effect of war on infectious disease, see Clara E. Councell, "War and Infectious Disease," *Public Health Reports* 56, no. 12 (March 21, 1941): 547–573.

2. Both the Black and white press in the United States reported on the sanitary conditions of hospitals during the Crimean War. See *The Provincial Freeman*, May 5, 1855; *Frederick Douglass' Paper*, August 31, 1855; London Friend, "Observations on the Crimean War," *Friends' Review; a Religious, Literary and Miscellaneous Journal (1847–1894)* 9, no. 7 (October 27, 1855): 101; "What the London Times Has Done," *United States Magazine of Science, Art, Manufactures, Agriculture, Commerce and Trade (1854–1856)* 2, no. 5 (1855): 174; "The Latest News from Europe," *Maine Farmer (1844–1900)* 23, no. 30 (July 19, 1855): 3; "Crimean Heroes and Trophies," *Frank Leslie's Weekly*, May 10, 1856; *The National Era*, December 7, 1854; *Godey's Lady's Book*, September 1855; *The Lily*, November 1, 1855; *Frank Leslie's Weekly*, March 1, 1856.

3. On the United States, see David Rosner, *A Once Charitable Enterprise: Hospitals and Health Care in Brooklyn and New York, 1885–1915* (Cambridge: Cambridge University Press, 1982); David J. Rothman, *The Discovery of the Asylum: Social Order and Disorder in the New Republic* (Boston: Little Brown, 1971.) On the Spanish Empire, see Pablo Gómez, *The Experiential Caribbean: Creating Knowledge and Healing in the Early Modern Atlantic* (Chapel Hill: University of North Carolina Press, 2017).

4. It was not until the twentieth century that the meaning of the hospital changed. See Rosner, *A Once Charitable Enterprise*; Paul Starr, *The Social Transformation of American Medicine: The Rise of a Sovereign Profession and the Making of a Vast Industry* (New York: Basic Books, 1982); Charles Rosenberg, *The Care of Strangers: The Rise of America's Hospital System* (Baltimore: Johns Hopkins University Press, 1995).

5. As agrarian-based household economies dwindled, a new population of so-called dependent people, such as elderly and disabled people and orphans, emerged. As wage-labor increased, fewer families could afford to take care of these people, and they ended up in hospitals, asylums, and orphanages. See David J. Rothman, *The Discovery of the Asylum: Social Order and Disorder in the New Republic* (Boston:

Little, Brown, 1971); Seth Rockman, *Scraping By: Wage Labor, Slavery, and Survival in Early Baltimore* (Baltimore: Johns Hopkins University Press, 2009); Downs, *Sick from Freedom.*

6. Medical care in these hospitals was often limited to food, shelter, and clothing. While some medical and government authorities distinguished hospitals for the sick from almshouses for the poor and asylums for the mentally disabled, many medical reformers described these institutions interchangeably (Downs, *Sick from Freedom,* 120–145). The poorly trained medical staffs often mistreated or neglected their charges. On asylums for mentally disabled patients, see Wendy Gonaver, *The Peculiar Institution and the Making of Modern Psychiatry, 1840–1880* (Chapel Hill: University of North Carolina Press, 2019); on asylums for the poor, Rockman, *Scraping By*; on orphanages for black children, Leslie Harris, *In the Shadow of Slavery: African Americans in New York City, 1626–1863* (Chicago: University of Chicago Press, 2004); on asylum management, Nancy Tomes, *A Generous Confidence: Thomas Story Kirkbride and the Art of Asylum-Keeping, 1840–1883* (Cambridge: Cambridge University Press, 1984).

7. On the abuse and violence that existed in nineteenth-century hospitals, particularly of the most oppressed people, see Harriet Washington, *Medical Apartheid: The Dark History of Medical Experimentation on Black Americans from Colonial Times to the Present* (New York: Harlem Moon, 2006), 67–69; Elaine G. Breslaw, *Lotions, Potions, Pills and Magic: Health Care in Early America* (New York: New York University Press, 2012), 145–147; Margaret Jones, "The Most Cruel and Revolting Crimes: The Treatment of the Mentally Ill in Mid-Nineteenth-Century Jamaica," *Journal of Caribbean History* 42, no. 2 (2008): 290–309; Emily Clark, "Mad Literature: Insane Asylums in Nineteenth Century America," *Arizona Journal of Interdisciplinary Studies* 4 (2015).

8. William Howard Russell, *The War: From the Landing at Gallipoli to the Death of Lord Raglan* (George Routledge & Co., 1855), 15, 63.

9. Quoted in Joseph J. Mathews, "The Father of War Correspondents," *Virginia Quarterly Review* 21, no. 1 (1945): 111–127.

10. Mark Bostridge, *Florence Nightingale: The Making of an Icon* (New York: Farrar, Straus and Giroux, 2008), 204 (quoting from an article by Thomas Chenery); Sue M. Goldie, *Florence Nightingale: Letters from the Crimea 1854–1856* (Manchester: Manchester University Press, 1997).

11. Goldie, *Florence Nightingale*, 15–21.

12. Goldie, *Florence Nightingale*, 3, 22.

13. Goldie, *Florence Nightingale*, 5.

14. Goldie, *Florence Nightingale*, 26.

15. *Times* (of London), November 13, 1854.

16. Florence Nightingale to Sydney Herbert, November 25, 1854, in Goldie, *Florence Nightingale*, 39.

17. Florence Nightingale to Mother, February 5, 1855, in Goldie, *Florence Nightingale*, 86–87.

18. Florence Nightingale to Sydney Herbert, February 19, 1855, in Goldie, *Florence Nightingale*, 93.

19. Hugo Small argues that Nightingale's efforts did not decrease the mortality rate; Small, *Florence Nightingale: Avenging Angel* (New York: St. Martin's, 1999). His analysis is disputed by Lynn McDonald in *Florence Nightingale: The Crimean War,* ed. Lynn McDonald, vol. 14 of *The Collected Works of Florence Nightingale* (Waterloo, ON: Wilfrid Laurier University Press, 2010), 32–36.

20. Florence Nightingale, *Notes on Hospitals: Being Two Papers Read before the National Association for the Promotion of Social Science, at Liverpool, in October, 1858. With Evidence Given to the Royal Commissioners on the State of the Army in 1857* (London: John W. Parker and Son, 1859), 40.

21. On rats in hospitals, see *Harper's Weekly,* May 5, 1860. On rats and reformers, see Starr, *The Social Transformation of American Medicine,* 155.

22. For some previous work on Nightingale's importance as a statistician and epidemiologist, see J. M. Keith, "Florence Nightingale: Statistician and Consultant Epidemiologist," *International Nursing Review* 35, no. 5, (1988): 147–150; L. R. C. Agnew, "Florence Nightingale: Statistician," *American Journal of Nursing* 58, no. 5 (1958): 664–665; Lynn McDonald, "Florence Nightingale and the Early Origins of Evidence-Based Nursing," *Evidence-Based Nursing* 4 (2001): 68–69; Lynn McDonald, "Florence Nightingale, Statistics, and the Crimean War," *Journal of the Royal Statistical Society Series A* 177, no. 3 (2014): 569–586; Edwin W. Kopf, "Florence Nightingale as Statistician," *Publications of the American Statistical Association* 15, no. 116 (1916): 388–404; Warren Winkelstein Jr., "Florence Nightingale: Founder of Modern Nursing and Hospital Epidemiology," *Epidemiology* 20, no. 2 (2009): 311; Christopher J. Gill and Gillian C. Gill, "Nightingale in Scutari: Her Legacy Reexamined," *Clinical Infectious Diseases* 40, no. 12 (2005): 1799–1805; D. Neuhauser, "Florence Nightingale Gets No Respect: As a Statistician That Is," *BMJ Quality & Safety* 12 (2003): 317.

23. Florence Nightingale, *Notes on Hospitals,* 3rd ed. (London: Longman, Green, Longman, Roberts and Green, 1863), 6–7.

24. Florence Nightingale, *Notes on Matters Affecting the Health, Efficiency, and Hospital Administration of the British Army, Founded Chiefly on the Experience of the Late War* (London: Printed by Harrison and Sons, 1858). For discussion and excerpts, see McDonald, ed., *Florence Nightingale: The Crimean War.*

25. According to Mark Bostridge, who has written an authoritative biography of Nightingale, the mortality rate at the hospital in Scutari was higher than in any other hospital when she first arrived. My concern, however, is not with her work in Scutari but with her advocacy of sanitary principles in her later writings. See Mark Bostridge, "Florence Nightingale: The Lady with the Lamp," BBC History, February 17, 2011, http://www.bbc.co.uk/history/british/victorians/nightingale_01.shtml.

26. "Netley Hospital," *Leeds Mercury,* August 21, 1858, no. 6837, https://cpb-ca-c1.wpmucdn.com/sites.uoguelph.ca/dist/3/30/files/2019/07/NEWSPAPE.pdf. Also see *Saturday Review,* August 28, 1858, 206–207.

27. Eduardo Faerstein and Warren Winkelstein Jr., "Adolphe Quetelet: Statistician and More," *Epidemiology* 12, no. 5 (2012): 762–763; Nathan Glazer, "The Rise of Social Science Research in Europe," in *The Science of Public Policy: Essential Readings in Policy Sciences I,* ed. Tadao Miyakawa (London: Routledge, 1999), 64.

28. Eileen Magnello, "Florence Nightingale: The Compassionate Statistician," +Plus Magazine, December 8, 2010, https://plus.maths.org/content/florence-nightingale -compassionate-statistician; Bostridge, *Florence Nightingale,* 306–315.

29. McDonald, "Florence Nightingale and the Early Origins of Evidence-Based Nursing."

30. Nightingale, *Notes on Hospitals* (1859), 3.

31. As quoted in Kopf, "Florence Nightingale as Statistician." I have relied on Kopf's description of the politics that led to the formation of the royal sanitary commission.

32. Nightingale, *Notes on Matters Affecting the Health . . . of the British Army,* 1–2.

33. Nightingale, *Notes on Matters Affecting the Health . . . of the British Army,* 89–90.

34. Nightingale, *Notes on Hospitals* (1859), 39–40.

35. Nightingale, *Notes on Matters Affecting the Health . . . of the British Army,* 492.

36. *Mortality of the British Army: At Home and Abroad, and during the Russian War, as Compared with the Mortality of the Civil Population in England* (London: Printed by Harrison and Sons, 1858), 12, table E. This booklet was reprinted from appendix 72 of the sanitary commission's report and is generally attributed to Nightingale. See Edward Tyas Cook, *The Life of Florence Nightingale,* vol. 2 (London: Macmillan, 1913), 381.

37. *Mortality of the British Army,* 5.

38. *Mortality of the British Army,* 16, table G. Nightingale suggested a revised classification of diseases in "Actual and Proposed Forms for Medical Statistics in the Army," *Notes on Matters Affecting the Health . . . of the British Army,* appendix 1 to section XI.

39. On the use of tables as new technology in reports, see McDonald, "Florence Nightingale and the Early Origins of Evidence-Based Nursing."

40. McDonald, "Florence Nightingale, Statistics, and the Crimean War"; Simon Rogers, "Florence Nightingale, Datajournalist: Information Has Always Been Beautiful," *The Guardian,* August 13, 2010.

41. Throughout the nineteenth century, certain religious orders specialized in nursing; on nurses in the Crimean War, see Maria Luddy, ed., *The Crimean Journals of the Sisters of Mercy, 1854–56* (Dublin: Four Courts Press, 2004); Mary Raphael Paradis, Edith Mary Hart, and Mary Judith O'Brien, "The Sisters of Mercy in the Crimean War: Lessons for Catholic Health Care," *Linacre Quarterly,* 84, no. 1 (2017): 29–43. Kaori Nagai argues that Nightingale erased the history of the Irish nuns, who applied what they had learned during the Irish famine to their work during the Crimean War. According to Nagai, these women threatened Nightingale's status as the "single female authority"; Nagai, "Florence Nightingale and the Irish Uncanny," *Feminist Review* 77 (2004): 26–45.

42. Nightingale, *Notes on Hospitals,* 3rd ed. (1863), 20–21.

43. Nightingale, *Notes on Hospitals,* 3rd ed. (1863), 10.

44. For a history of preventive medicine, see Daniel M. Becker, "History of Preventive Medicine," in *Prevention in Clinical Practice,* ed. Daniel M. Becker and Laurence B. Gardner (Boston: Springer, 1988), 13–21.

45. Nightingale, *Notes on Hospitals,* 3rd ed. (1863), 48–49.

46. "Brucellosis," Centers for Disease Control and Prevention, https://www.cdc.gov /brucellosis/index.html.

47. Gérard Vallée and Lynn McDonald, eds., *Florence Nightingale on Health in India,* vol. 9 of *The Collected Works of Florence Nightingale* (Waterloo, ON: Wilfrid Laurier University Press, 2006), xiii.

48. "Something of What Florence Nightingale Has Done and Is Doing," *St. James's Magazine* 1 (April 1861): 33; "The British Army and Miss Nightingale," *Medical and Surgical Reporter* 11, no. 18 (April 30, 1864): 267.

49. Nightingale, *Notes on Hospitals,* 3rd ed. (1863), 11–18.

50. William Dalrymple, "The East India Company: The Original Corporate Raiders," *The Guardian,* March 4, 2015.

51. Edwin H. H. Collen, "The Indian Army," in *The Empire and the Century: A Series of Essays on Imperial Problems and Possibilities by Various Writers* (New York: Dutton, 1905), 670.

52. Vallée and McDonald, eds., *Florence Nightingale on Health in India,* 12–14, 45–47.

53. Florence Nightingale, postscript to *Note on Matters Affecting the Health . . . of the British Army,* 656–667; Vallée and McDonald, eds., *Florence Nightingale on Health in India,* 46–47.

54. For an overview of how Nightingale's opinion of India evolved from an imperialist position to one of arguing for Indians to gain more power, see Jharna Gourlay, *Florence Nightingale and the Health of the Raj* (Aldershot, UK: Ashgate, 2003). On British derogatory attitudes toward Indian medicine and disregard of Indian healing and medical practices, see David Arnold, *Colonizing the Body: State Medicine and Epidemic Disease in Nineteenth-Century India* (Berkeley: University of California Press, 1993). On Indian medical practices, see Projit Bihari Mukharji, *Nationalizing the Body: The Medical Market, Print, and Daktari Medicine* (London: Anthem Press, 2009).

55. Quoted in Vallée and McDonald, eds., *Florence Nightingale on Health in India,* 7.

56. Vallée and McDonald, eds., *Florence Nightingale on Health in India,* 8.

57. Vallée and McDonald, eds., *Florence Nightingale on Health in India,* 8–9.

58. Vallée and McDonald, eds., *Florence Nightingale on Health in India,* 27.

59. Florence Nightingale, *Observations on the Evidence Contained in the Stational Reports Submitted to Her by the Royal Commission on the Sanitary State of the Army in India* (London: Edward Stanford, 1863), 17.

60. Vallée and McDonald, eds., *Florence Nightingale on Health in India,* 15.

61. In New York City, for example, the Board of Health created a special sanitary committee to respond to the outbreak of cholera. Often histories of these organizations describe them as outgrowths of municipal government and divorce them from larger imperial and global transformations. While local governments did serve as the origin for some of these efforts, such as Chadwick's studies in London, imperialism popularized the use of bureaucracy and the practice of putting together a board of experts to oversee the health of a region. On such associations in New York City, see Charles Rosenberg, *The Cholera Years: The United States in 1832, 1849 and 1866,* 2nd ed. (Chicago: University of Chicago, 1987), 108–109.

62. Gyan Prakash, *Another Reason: Science and the Imagination of Modern India* (Princeton: Princeton University Press, 1999), 132–135.

63. In England at this time, statistics also became an important tool for investigating social problems and conveying these issues to a broad public. See Oz Frankel, *States of Inquiry: Social Investigations and Print Culture in Nineteenth-Century Britain and the United States* (Baltimore: Johns Hopkins University Press, 2006).

64. Between roughly 1700 and 1827, scholars drew on statistical knowledge to study astronomy and the atmosphere. Prince Albert's mentor Adolphe Quetelet ushered in the use of statistics to create a "calculus of probabilities" to be applied to the social sciences in the nineteenth century. Quetelet and Nightingale were friends. For a history of statistics, see Stephen M. Stigler, *The History of Statistics: The Measurement of Uncertainty before 1900* (Cambridge: Harvard University Press, 1990). On Nightingale's friendship with Quetelet, see Kopf, "Florence Nightingale as Statistician."

65. Kopf, "Florence Nightingale as Statistician," 396. Kopf notes that Nightingale's plan for collection of uniform hospital statistics was difficult to implement and "was never generally successful over any considerable period of time."

66. As quoted in David J. Spiegelhalter, "Surgical Audit: Statistical Lessons from Nightingale and Codman," *Journal of the Royal Statistical Society, Series A* 162, no. 1 (1999), 47. See also James T. Hammack, "Report to the Statistical Society on the Proceedings of the Fourth Session of the International Statistical Congress, Held in London, July, 1860," *Journal of the Statistical Society of London* 24, no. 1 (1861): 6.

67. Kopf, "Florence Nightingale as Statistician."

68. Florence Nightingale to William Farr, April 28, 1860, in McDonald, "Florence Nightingale and the Early Origins of Evidence-Based Nursing."

69. McDonald, "Florence Nightingale and the Early Origins of Evidence-Based Nursing."

70. Kopf, "Florence Nightingale as Statistician, " 397.

71. Steve M. Blevins and Michael S. Bronze, "Robert Koch and the 'Golden Age' of Bacteriology," *International Journal of Infectious Diseases* 14, no. 9 (2010): e744–e751; "Who First Discovered *Vibrio cholera*?" UCLA Department of Epidemiology, http://www.ph.ucla.edu/epi/snow/firstdiscoveredcholera.html.

72. Vallée and McDonald, eds., *Florence Nightingale on Health in India*, 863–65.

73. Nightingale's views on the spread of infectious disease are collected in *Florence Nightingale on Public Health Care*, ed. Lynn McDonald, vol. 6 of *The Collected Works of Florence Nightingale* (Waterloo, ON: Wilfrid Laurier University Press, 2004). Nightingale's ideas about the social and environmental conditions that cause sickness dovetail with the current thinking of medical anthropologists. See, for example, Paul Farmer, *Infections and Inequalities: The Modern Plagues* (Berkeley: University of California Press, 1999).

74. Other British epidemiologists, like Edward Ballard, John Netten Radcliffe, George Buchanan, and Richard Thorne, accepted the germ theory and argued for an environmentalist view. I am grateful to Jacob Steere-Williams for this insight.

75. Florence Nightingale to Thomas Gillham Hewlett, July 27, 1883, in Vallée and McDonald, eds., *Florence Nightingale on Health in India*, 921.

76. Nightingale to Hewlett, 921.

77. Florence Nightingale to Lord Dufferin, November 4, 1886, in Vallée and McDonald, eds., *Florence Nightingale on Health in India*, 925.

78. Nightingale to Dufferin, 925.

79. Nightingale, *Observations on the Evidence Contained in the Stational Reports*, 51. She delivered a paper, for example, at the National Association for the Promotion of Social Science in Edinburgh in 1863, claiming that "native races" had a tendency to disappear following contact with "the influences of civilization"; Kopf, "Florence Nightingale as Statistician," 399.

80. While many scholars recognize that imperialism provided the laboratory for the European Enlightenment and contributed to the formation of many disciplines, I am explaining how this played out in the specific subfield of epidemiology and public health. On the former, see Sankar Muthu, *Empire and Modern Political Thought* (Cambridge: Cambridge University Press, 2012).

81. Richard J. Evans, *Death in Hamburg: Society and Politics in the Cholera Years, 1830–1910* (Oxford: Oxford University Press, 1987), 268–269.

82. Evans, *Death in Hamburg*, 671–707.

83. Mariko Ogawa, "Uneasy Bedfellows: Science and Politics in the Refutation of Koch's Bacterial Theory of Cholera," *Bulletin of the History of Medicine* 74, no. 4 (2000): 705.

84. David Arnold notes that Indians of "almost every caste and class vehemently opposed postmortems." Arnold, *Colonizing the Body: State Medicine and Epidemic Disease in Nineteenth-Century India* (Berkeley: University of California Press, 1993), 53.

85. Gill and Gill, "Nightingale in Scutari."

6. From Benevolence to Bigotry

1. Leslie Schwalm has recently published an article on how the US Sanitary Commission supported racial essentialism and used its allegedly "scientific" investigations to buttress racial hierarchies. In this chapter I put the USSC in a broader global context and show how USSC codified race as a key if atavistic metric in the development of epidemiology. Leslie A. Schwalm, "A Body of 'Truly Scientific Work': The U.S. Sanitary Commission and the Elaboration of Race in the Civil War Era," *Journal of the Civil War Era* 8, no. 4 (2018): 647–676.

2. Scholars Barbara Fields and Dorothy Roberts have emphasized that race is not a biological category, yet for many in the medical profession, both in the past and the present, it is treated like one. Barbara J. Fields, "Slavery, Race, and Ideology in the United States of America," *New Left Review* 181 (June 1990), 95–118; Dorothy Roberts, *Fatal Invention: How Science, Politics, and Big Business Re-Create Race in the Twenty-First Century* (New York: New Press, 2011).

3. Julia Boyd, "Florence Nightingale and Elizabeth Blackwell," *Lancet* 373, no. 9674 (2009): P1516–1517.

4. Judith Ann Giesberg, *Civil War Sisterhood: The U.S. Sanitary Commission and Women's Politics in Transition* (Boston: Northeastern University Press, 2000), 33.

5. Giesberg, *Civil War Sisterhood*, 35.

6. Giesberg, *Civil War Sisterhood*, 46.

7. "Order creating the United States Sanitary Commission, signed and approved by President Lincoln on June 13, 1861. Countersigned by Simon Cameron, Secretary of War," New York Public Library Digital Collections, https://digitalcollections.nypl .org/items/510d47e4-5a05-a3d9-e040-e00a18064a99; *The United States Sanitary Commission: A Sketch of Its Purposes and Its Work* (Little, Brown and Company, 1863); "obtrusive" quote from Jeanie Attie, *Patriotic Toil: Northern Women and the American Civil War* (Ithaca, NY: Cornell University Press, 1998), 53.

8. On women's work in the USSC, see Giesberg, *Civil War Sisterhood,* and Attie, *Patriotic Toil.*

9. There is not yet an epidemiological analysis of the Civil War that tracks how population movements and the destruction of natural resources spurred epidemics, but see Urmi Willoughby on how shifts in population owing to the slave trade, the conversion of land into plantations, and other factors led to a yellow fever outbreak in Louisiana, and Megan Nelson on how armies destroyed nature by cutting down countless numbers of trees. Urmi Engineer Willoughby, *Yellow Fever, Race, and Ecology in Nineteenth Century New Orleans* (Baton Rouge: Louisiana State University Press, 2014); Megan Kate Nelson, *Ruin Nation: Destruction and the American Civil War* (Athens: University of Georgia Press, 2012).

10. Louis C. Duncan, "The Medical Department of the United States Army in the Civil War—The Battle of Bull Run," *The Military Surgeon* 30, no. 6 (1912), 644–668, on 667.

11. Duncan, "Medical Department of the United States Army," 644. Also see Jim Downs, *Sick from Freedom: African American Illness and Suffering during the Civil War and Reconstruction* (New York: Oxford University Press, 2012), 28–31.

12. Duncan, "Medical Department of the United States Army," 665–666.

13. *The Sanitary Commission of the United States Army: A Succinct Narrative of Its Works and Purposes* (New York: United States Sanitary Commission, 1864), 14.

14. Charles J. Stillé, *History of the United States Sanitary Commission: Being the General Report of Its Work during the War of the Rebellion* (Philadelphia: J. B. Lippincott, 1866), 27.

15. Stillé, *History of the United States Sanitary Commission,* 32.

16. US Sanitary Commission, *Rules for Preserving the Health of the Soldier* (Washington, DC, 1861). On changing practices of cleanliness, see Kathleen M. Brown, *Foul Bodies: Cleanliness in Early American Society* (New Haven, CT: Yale University Press, 2009).

17. US Sanitary Commission, *Rules for Preserving the Health of the Soldier,* 6–7.

18. On hospitals, see, for example, W. Gill Wylie, *Hospitals: Their History, Organization, and Construction* (New York: D. Appleton, 1877).

19. US Sanitary Commission, *Rules for Preserving the Health of the Soldier,* 8.

20. Stillé, *History of the United States Sanitary Commission,* 27–32.

21. In the 1840s, Josiah Nott suggested that there was a relationship between mosquitoes and yellow fever, but I have not seen any evidence in the USSC records that sanitary reformers in the North were familiar with this theory. Josiah C. Nott, "Yellow Fever Contrasted with Bilious Fever—Reasons for Believing It a Disease Sui Generis—Its Mode of Propagation—Remote Cause: Probable Insect or

Animalcular Origin, &c.," *New Orleans Medical and Surgical Journal* 4 (1848): 563–601.

22. Gail Bederman showed that at the turn of the twentieth century, many elite men rebelled against efforts by women to implement manners, etiquette, and other domestic sensibilities. I imagine that during the Civil War, a similar phenomenon took place and many military officials and troops ignored sanitary regulations, particularly because they were espoused by women. Gail Bederman, *Manliness and Civilization: A Cultural History of Gender and Race, 1880–1917* (Chicago: University of Chicago Press, 1995).

23. Shauna Devine, *Learning from the Wounded: The Civil War and the Rise of American Medical Science* (Chapel Hill: University of North Carolina Press, 2014), 1; Downs, *Sick from Freedom,* 4; Downs, "#BlackLivesMatter: Toward an Algorithm of Black Suffering during the Civil War and Reconstruction," *J19: The Journal of Nineteenth-Century Americanists* 4, no. 1 (2016): 198–206.

24. Stillé's *History of the United States Sanitary Commission* offers a comprehensive overview of the USSC's work, the role specific doctors played and, most of all, how the USSC taught soldiers and volunteers about the urgency of promoting sanitary environments.

25. *The Liberator,* May 30, 1862.

26. On the USSC seal, see Giesberg, *Civil War Sisterhood,* vii–x.

27. Jennifer L. Morgan, *Laboring Women: Reproduction and Gender in New World Slavery* (Philadelphia: University of Pennsylvania Press, 2004); Rana A. Hogarth, *Medicalizing Blackness: Making Racial Difference in the Atlantic World* (Chapel Hill: University of North Carolina Press, 2017).

28. Downs, *Sick from Freedom.*

29. Susan P. Waide and Valerie Wingfield, "United States Sanitary Commission Records, 1861–1878, MssCol 3101," New York Public Library, Manuscripts and Archives Division, January 2006, 10, https://www.nypl.org/sites/default/files /archivalcollections/pdf/ussc.pdf.

30. Waide and Wingfield, "United States Sanitary Commission Records, . . . MssCol 3101," 10.

31. Margaret Humphreys, *Intensely Human: The Health of the Black Soldier in the American Civil War* (Baltimore: Johns Hopkins University Press, 2008), 8.

32. Although distinctions were made among immigrants, particularly Irish immigrants, these ideas do not span across generations in the same way that they do for Black people. Based on the data that the USSC collected, Benjamin Gould published a study that outlined how Black soldiers differed from others who served in the military. Decades later, Frederick Hoffman drew on this research to propagate ideas about racial inferiority that informed the late-nineteenth-century discourse on eugenics. Benjamin Apthorp Gould, *Investigations in the Military and Anthropological Statistics of American Soldiers* (New York: US Sanitary Commission, 1869), 347–348, 391, 479; Frederick L. Hoffman, *Race Traits and Tendencies of the American Negro* (New York: Macmillan, for the American Economic Association, 1896), 70–71, 150–151, 162–168, 183–185.

33. A copy of the questionnaire can be found in the correspondence of USSC officials: Benjamin Woodward to Elisha Harris, August 20, 1863, Memphis, TN, series 1: Medical Committee Archives, 1861–1866, United State Sanitary Commission Records, New York Public Library (hereafter, USSC Records). On women reformers committed to antiracism as part of their larger activist agendas, see Nancy A. Hewitt, *Women's Activism and Social Change: Rochester, 1822–1872* (Ithaca, NY: Cornell University Press, 1984); Nancy A. Hewitt, *Radical Friend: Amy Kirby Post and Her Activist Worlds* (Chapel Hill: University of North Carolina Press, 2018).

34. Quoted in Humphreys, *Intensely Human*, 33.

35. Josiah Nott, "The Mulatto a Hybrid: Probable Extermination of the Two Races If the Whites and Blacks Are Allowed to Intermarry," *Boston Medical and Surgical Journal* 29, no. 2 (August 16, 1843): 29–32.

36. Ira Russell, "Hygienic and Medical Notes and Report on Hospital L'Ouverture," reel 3, frame 282, USSC Records.

37. Nott, "The Mulatto a Hybrid," 30.

38. Nott, "Yellow Fever Contrasted with Bilious Fever."

39. Walter Johnson, *Soul by Soul: Inside the Antebellum Slave Market* (Cambridge: Harvard University Press, 1999), 155.

40. Walter Johnson also finds evidence for these practices beyond the popular representation in literature. Johnson, *Soul by Soul*, 154.

41. On systems of racial classification in the United States in the context of the Atlantic World, see Jim Downs, "Her Life, My Past: Rosina Downs and the Proliferation of Racial Categories after the American Civil War," in *Storytelling, History, and the Postmodern South*, ed. Jason Philips (Baton Rouge: Louisiana State University Press, 2013).

42. Josiah Nott was the leading advocate of this theory. J. C. Nott, "Hybridity of Animals, Viewed in Connection with the Natural History of Mankind," in J. C. Nott and G. R. Gliddon, *Types of Mankind: or, Ethnological Researches, Based upon the Ancient Monuments, Paintings, Sculptures, and Crania or Races and upon their Natural, Geographical, Philological, and Biblical History,* 2nd ed. (Philadelphia: Lippincott, Grambo & Co., 1854), 397–399.

43. Gould, *Investigations in the Military and Anthropological Statistics of American Soldiers,* 347–353, 471, 478. For an overview of scientific discussions on the intersection of race and lung capacity, in which Gould played a prominent role, see Lundy Braun, *Breathing Race into the Machine: The Surprising Career of the Spirometer from Plantation to Genetics* (Minneapolis: University of Minnesota Press, 2014).

44. Melissa N. Stein, *Measuring Manhood: Race and the Science of Masculinity* (Minneapolis: University of Minnesota Press, 2015), 106–107.

45. Humphreys, *Intensely Human*, 34.

46. Humphreys, *Intensely Human*, 35.

47. Russell, "Hygienic and Medical Notes and Report on Hospital L'Ouverture."

48. George Andrew to Elisha Harris, July 17, 1865, LaPorte, Indiana, reel 1, frame 408, USSC Records.

49. Benjamin Woodward, "Report on the Diseases of the Colored Troops," reel 2, frame 867, USSC Records. The federal government's decision to study specific diseases among Black people was also part of the Freedmen's Bureau, which treated formerly enslaved people who were not part of the army and were outside the USSC's reach. See, for example, Robert Reyburn, who challenged the USSC's arguments about the cause of disease among Black people, but whose publication, nonetheless, reified racial categories by using medicine as a justification to examine disease specific to freedpeople. Reyburn, *Types of Disease among Freed People of the United States* (Washington, DC: Gibson Bros., 1891). On the Freedmen's Bureau Medical Division and Reyburn, see Downs, *Sick from Freedom*.

50. Humphreys, *Intensely Human*, 34.

51. On the history of Black people and tuberculosis, see Samuel Kelton Roberts, *Infectious Fear: Politics, Disease and the Health Effects of Segregation* (Chapel Hill: University of North Carolina Press, 2009); Braun, *Breathing Race into the Machine*.

52. Woodward, "Report on the Diseases of the Colored Troops."

53. Outside of medical reports, in writings by Northern abolitionists who worked as benevolent reformers during the postwar period, there are references to Christianity, which also appear in the education records. See *Freedmen's Record* 1, no. 10 (October 1865), 160. On Christianity in schools, see Hilary Green, *Educational Reconstruction: African American Schools in the Urban South, 1865–1890* (New York: Fordham University Press, 2016), 22, 95.

54. On how enslaved people's ideas about health and medicine drew on Indigenous and European traditions, see Sharla Fett, *Working Cures: Healing, Health and Power on Southern Slave Plantations* (Chapel Hill: University of North Carolina Press, 2002).

55. Joseph R. Smith, "Sanitary Report of the Department of Arkansas for the year 1864," MS C 126, Historical Collection, National Library of Medicine, Bethesda, MD, as quoted in Humphreys, *Intensely Human*, 15, 165.

56. Doctors in New Orleans traced the spread of yellow fever to poor sanitary conditions. See, for example, Erasmus D. Fenner, *History of the Epidemic Yellow Fever: At New Orleans, La., in 1853* (New York: Hall, Clayton, 1854).

57. I am building on Margaret Humphreys's research on Ira Russell and his work. See Humphreys, *Intensely Human*, xi.

58. Humphreys, *Intensely Human*, 46, 51.

59. Humphreys, *Intensely Human*, x–xiii.

60. Humphreys, *Intensely Human*, 9. Few doctors during the Civil War recognized that emancipation caused massive medical crises. See Downs, *Sick from Freedom*.

61. Humphreys, *Intensely Human*, x, 51, 100–102.

62. Samuel George Morton, *Crania Americana; or, A Comparative View of the Skulls of Various Aboriginal Nations of North and South America* (Philadelphia: J. Dobson, 1839), 7; also see 87–88.

63. "Morton's Later Career and Craniology," Morton Collection, University of Pennsylvania Museum of Archeology and Anthropology, https://www.penn.museum/sites/morton/life.php.

64. On the Black intellectual response to Morton, see Britt Russert, *Fugitive Science: Empiricism and Freedom in Early African American Culture* (New York: NYU Press, 2017).

65. On Morton, see Ann Fabian, *The Skull Collectors: Race, Science, and America's Unburied Dead* (Chicago: University of Chicago Press, 2010), 36–43. On the number of autopsies, see Humphreys, *Intensely Human,* 100.

66. Michal Sappol, *A Traffic of Dead Bodies: Anatomy and Embedded Social Identity in Nineteenth-Century America* (Princeton, NJ: Princeton University Press, 2004); Daina Ramey Berry, *The Price for Their Pound of Flesh: The Value of the Enslaved, from Womb to Grave, in the Building of a Nation* (Boston: Beacon, 2018).

67. Barbara and Karen Fields developed the term "racecraft" to explain how those in power contribute to racial ideology by developing ideas or theories that support inferiority. If people were not actively creating an ideology about "race," it would not exist; like witchcraft, race gains power based on the extent to which people believe in its potency. Karen E. Fields and Barbara J. Fields, *Racecraft: The Soul of Inequality in America* (London: Verso, 2016).

68. Hygienic and Medical Notes, Questions and Answers. Dr. Russell (no date), reel 5, frame 284, USSC Records.

69. Ira Russell, January 31, 1864, folder 2: papers: scan 21–22, Ira Russell Papers #4440, Southern Historical Collection, Wilson Library, University of North Carolina at Chapel Hill (hereafter, Russell Papers).

70. I have not seen any of his correspondence from before the war. His archived letters come mostly from his wartime service and were preserved, in part, by his son (in the Russell Papers collection).

71. Russell Papers; Ira Russell Letters, 1862–1863, Collection MC 581, Special Collections Department, University of Arkansas Libraries, Fayetteville, AR. On his later influence, see Humphreys, *Intensely Human,* xiii. On citation by racialist popular thinkers after the Civil War, see Hoffman, *Race Traits and Tendencies of the American Negro,* 159.

72. "In Memory of Dr. Elisha Harris," *Public Health Papers and Reports* 10 (1884): 509–510. On the influence of Civil War doctors on American medicine, see Shauna Devine, *Learning from the Wounded: The Civil War and the Rise of American Medical Science* (Chapel Hill: University of North Carolina Press, 2014).

73. According to the historian Lundy Braun, Gould's analysis influenced Darwin's "scientific arguments for a hierarchy of difference." Braun, *Breathing Race into the Machine,* 41.

74. On how Black people fought against laws that deemed them inferior, see Martha S. Jones, *Birthright Citizens: A History of Race and Rights in Antebellum America* (Cambridge: Cambridge University Press, 2018).

75. See, for example, Austin Flint, *Contributions Relating to the Causation and Prevention of Disease, and to Camp Diseases; Together with a Report of the Diseases, etc., among the Prisoners at Andersonville, Ga.* (New York: U.S. Sanitary Commission, 1867) 5, 170, 290, 319, 333, 664.

76. Roberts, *Fatal Invention.*

77. Mary A. Livermore, *My Story of the War: A Woman's Narrative of Four Years Personal Experience* (Hartford: A. D. Worthington, 1890); Laura S. Haviland, *A Woman's Life-Work; Labors and Experiences* (Chicago: Waite, 1887); Katherine Prescott Wormeley, *The Other Side of the War with the Army of the Potomac* (Boston: Ticknor and Fields, 1889).

7. "Sing, Unburied, Sing"

The chapter title comes from Jesmyn Ward's award-winning novel *Sing, Unburied, Sing* (New York: Scribner, 2017).

1. On slaveholders assigning a price to enslaved people based on their age and gender, see Daina Ramey Berry, *The Price for Their Pound of Flesh: The Value of the Enslaved, from Womb to Grave, in the Building of a Nation* (Boston: Beacon Press, 2017).

2. I am drawing on Robert D. Hicks's description of the wartime procedure of vaccination as well as his meticulous description of Civil War doctors' vaccination kits. Hicks, "Scabrous Matters: Spurious Vaccinations in the Confederacy," in *War Matters: Material Culture in the Civil War Era,* ed. Joan E. Cashin (Chapel Hill: University of North Carolina Press), 126.

3. Suzanne Krebsbach, "The Great Charlestown Smallpox Epidemic of 1760," *South Carolina Historical Magazine* 97, no. 1 (1996): 30–37; Alan D. Watson. "Combating Contagion: Smallpox and the Protection of Public Health in North Carolina, 1750 to 1825," *North Carolina Historical Review* 90, no. 1 (2013): 26–48; Elizabeth Fenn, *Pox Americana: The Great Smallpox Epidemic of 1775–82* (New York: Hill and Wang, 2001), 39–42.

4. On the causes of, responses to, and effects of the smallpox epidemics during the Civil War, see Jim Downs, *Sick from Freedom: African American Illness and Suffering during the Civil War and Reconstruction* (New York: Oxford University Press, 2012). On the abrupt dislocation that the war engendered throughout the South among soldiers, former bondspeople, and civilians, see Yael Sternhell, *Routes of War: The World of Movement in the Confederate South* (Cambridge, MA: Harvard University Press, 2012).

5. Hicks, "Scabrous Matters," 128–130.

6. Historians agree that variolation was developed in Asia and Africa long before it arrived in America. The exact date is not known; some have hypothesized that it originated as early as 1000 BCE. See, for example, Stefan Riedel, "Edward Jenner and the History of Smallpox and Vaccination," *Baylor University Medical Center Proceedings* 18, no. 1 (2005): 21–25.

7. Margot Minardi, "The Boston Inoculation Controversy of 1721–1722: An Incident in the History of Race," *William and Mary Quarterly* 61, no. 1 (2004): 47–76; Harriet Washington, *Medical Apartheid: The Dark History of Medical Experimentation on Black Americans from Colonial Times to the Present* (New York: Anchor, 2008), 70–73; "How an African Slave Helped Boston Fight Smallpox," *Boston Globe,* October 17, 2014; Benjamin Waterhouse, *A Prospect of Exterminating the Small-Pox: Being the History of the Variolæ Vaccinæ, or Kine-pox, Commonly Called the*

Cow-Pox; As It Has Appeared in England: With an Account of a Series of Inocula-
tions Performed for the Kine-Pox, in Massachusetts ([Cambridge, MA]: Cambridge
Press, 1800).

8. Minardi, "Boston Inoculation Controversy"; see also Kelly Wisecup, "African
    Medical Knowledge, the Plain Style, and Satire in the 1721 Boston Inoculation
    Controversy," *Early American Literature* 46, no. 1 (2011): 25–50.

9. Minardi, "Boston Inoculation Controversy."

10. On the flow of information from Africa to the New World, in particular the
    cross-pollination of medical knowledge as a result of slavery, see Pablo F. Gómez,
    The Experiential Caribbean: Creating Knowledge and Healing in the Early Modern
    Atlantic (Chapel Hill: University of North Carolina, 2017); Sharla M. Fett, *Working*
    Cures: Healing, Health, and Power on Southern Slave Plantations (Chapel Hill:
    University of North Carolina Press, 2002).

11. Hicks, "Scabrous Matters," 131–132.

12. Joseph Jones, *Researches upon "Spurious Vaccination"; or The Abnormal Phe-*
    nomena Accompanying and Following Vaccination in the Confederate Army, during
    the Recent American Civil War, 1861–1865 (Nashville: University Medical Press,
    1867), 85.

13. Lydia Murdoch, "Carrying the Pox: The Use of Children and Ideals of Childhood
    in Early British and Imperial Campaigns against Smallpox," *Journal of Social*
    History 48, no. 3 (2015): 511–535.

14. For mention of obtaining vaccine matter from infants, see Jones, *Researches upon*
    "Spurious Vaccination," 27, 76, 86.

15. For images of nineteenth-century lancets used for vaccination, see "Vaccination
    Instruments," History of Vaccines, College of Physicians of Pennsylvania,
    https://www.historyofvaccines.org/index.php/content/vaccination-instruments.

16. In order to develop this sketch, I drew on both primary and secondary evidence
    that provided details about the inoculation process. On the importance of the
    eighth day, see Jones, *Researches upon "Spurious Vaccination,"* 50, 55; Hicks,
    "Scabrous Matters," 128–130.

17. For more on how the archive has been complicit with studies of slavery and capitalism
    and why historians need to do more to resist the categories that slaveholders used to
    define slavery, see Jim Downs, "When the Present Is Past: Writing the History of
    Sexuality and Slavery," in *Sexuality and Slavery: Reclaiming Intimate Histories in*
    the Americas, ed. Daina Ramey Berry and Leslie M. Harris (Athens: University of
    Georgia Press, 2018).

18. Daina Ramey Berry, in *The Price for Their Pound of Flesh,* offers one of the few
    accounts of how the price of enslaved people varied by age and race. Children were
    often not counted as laborers, and infants may have been disregarded because of
    their lack of economic value. In the face of widespread epidemics, however, infants
    became enormously valuable. Wendy Warren, "'Thrown upon the World': Valuing
    Infants in the Eighteenth-Century North American Slave Market," *Slavery and*
    Abolition 39, no. 4 (2018): 623–641.

19. C. Michele Thompson, *Vietnamese Traditional Medicine: A Social History*
    (Singapore: NUS Press, 2015), 28–43; Murdoch, "Carrying the Pox."

20. Jones, *Researches upon "Spurious Vaccination,"* 72–73.

21. Frank Ramsey, in Jones, *Researches upon "Spurious Vaccination,"* 94.

22. Paul Eve, in Jones, *Researches upon "Spurious Vaccination,"* 90.

23. Elisha Harris, "Vaccination in the Army—Observations on the Normal and Morbid Results of Vaccination and Revaccination during the War, and on Spurious Vaccination," in *Contributions Relating to the Causation and Prevention of Disease, and to Camp Diseases; Together with a Report of the Diseases, etc., among the Prisoners at Andersonville, Ga*, ed. Austin Flint (New York: US Sanitary Commission, 1867), 143–145.

24. Harris, "Vaccination in the Army," 143.

25. Harris, "Vaccination in the Army," 144.

26. Barbara J. Fields, "Who Freed the Slaves?" in *The Civil War: An Illustrated History*, ed. Geoffrey C. Ward (New York: Knopf, 1990).

27. Quoted in Sharon Romeo, *Gender and the Jubilee: Black Freedom and the Reconstruction of Citizenship in Civil War Missouri* (Athens: University of Georgia Press, 2016), 35.

28. Downs, *Sick from Freedom;* Romeo, *Gender and the Jubilee.*

29. Russell, quoted in Harris, "Vaccination in the Army," 145, 148.

30. Ira Russell, January 31, 1864, folder 2: papers: scan 23–27, Ira Russell Papers #4440, Southern Historical Collection, Wilson Library, University of North Carolina at Chapel Hill.

31. For more on the vulnerable conditions of freedpeople during the Civil War and Reconstruction, as well as the ways in which the organization of the labor force left children especially susceptible to exploitation and abuse, see Downs, *Sick from Freedom.*

32. This is not pronounced or even mentioned in the otherwise excellent historiography. On slavery and medicine, see Todd L. Savitt, *Medicine and Slavery: The Diseases and Health Care of Blacks in Antebellum Virginia* (Urbana: University of Illinois, 1978); Fett, *Working Cures;* Marie Jenkins Schwartz, *Birthing a Slave: Motherhood and Medicine in the Antebellum South* (Cambridge: Harvard University Press, 2010); Stephen C. Kenny, "'A Dictate of Both Interest and Mercy'? Slave Hospitals in the Antebellum South," *Journal of the History of Medicine and Allied Sciences* 65, no. 1 (2010): 1–47; Deirdre Cooper Owens, *Medical Bondage: Race, Gender, and the Origins of American Gynecology* (Athens: University of Georgia Press, 2017). On the reporting of violence and its history during slavery, see Walter Johnson, *Soul by Soul: Life inside the Antebellum Slave Market* (Cambridge: Harvard University Press, 1999); Edward E. Baptist, *The Half Has Never Been Told: Slavery and the Making of American Capitalism* (New York: Basic Books, 2014); and Marisa J. Fuentes, *Dispossessed Lives: Enslaved Women, Violence, and the Archive* (Philadelphia: University of Pennsylvania Press, 2016).

33. Circular No. 2, Surgeon General's Office, February 6, 1864, as quoted in Carol Cranmer Green, "Chimborazo Hospital: A Description and Evaluation of the Confederacy's Largest Hospital" (PhD dissertation, Texas Tech University, 1999), 283. See also H. H. Cunningham, *Doctors in Gray: The Confederate Medical Service* (1958; Baton Rouge: Louisiana State University Press, 1986), 201; Hicks, "Scabrous Matters," 139.

34. Joseph Jones, *Contagious and Infectious Diseases, Measures for Their Prevention and Arrest . . . ,* Circular No. 2, Board of Health of the State of Louisiana (Baton Rouge: Leon Jastremski, 1884), 282. For other references to Bolton, see Charles Smart, *The Medical and Surgical History of the War of the Rebellion,* part 3, vol. 1: *Medical History* (Washington, DC: Government Printing Office, 1888), 645–646; Donald R. Hopkins, *The Greatest Killer: Smallpox in History* (Chicago: University of Chicago Press, 2002), 276; Hicks, "Scabrous Matters," 123.

35. Harris, "Vaccination in the Army," 157.

36. While my concern is with medicine as a site of reconciliation, other historians have written about reconciliation in the areas of politics, literature, and social issues. See David Blight, *Race and Reunion: The Civil War in American Memory* (Cambridge: Harvard University Press, 2001); Caroline E. Janney, *Remembering the Civil War: Reunion and the Limits of Reconciliation* (Chapel Hill: University of North Carolina Press, 2013); Nina Silber, *The Romance of Reunion: Northerners and the South, 1865–1900* (Chapel Hill: University of North Carolina Press, 1993).

37. Jane Zimmerman, "The Formative Years of the North Carolina Board of Health," *North Carolina Historical Review* 21, no. 1 (1944), 3. For more on how emancipation gave way to massive dislocation that led to the spread of infectious disease, particularly smallpox, see Downs, *Sick from Freedom.*

38. Howard A. Kelly and Walter A. Burrage, *American Medical Biographies* (Baltimore: Norman, Remington, 1920), 1259–1260.

39. Thomas F. Wood, "Vaccination: A Consideration of Some Points as to the Identity of Variola and Vaccina," *Chicago Medical Journal and Examiner* 43, no. 4 (1881): 347–356; John Joseph Buder, "Letters of Henry Austin Martin: The Vaccination Correspondence to Thomas Fanning Wood, 1877–1883" (master's thesis, University of Texas at Austin, 1991).

40. Wood, "Vaccination," 352.

41. Zimmerman, "Formative Years."

42. James O. Breeden, "Joseph Jones and Confederate Medical History," *Georgia Historical Quarterly* 54, no. 3 (1970): 357–380.

43. Joseph Jones to S. P. Moore, June 28, 1863, in "Biographical Sketch of Joseph Jones," *Physicians and Surgeons of America,* ed. Irving Watson (Concord, NH: Republican Press Association, 1896), 593–597, available at https://collections.nlm .nih.gov/pdf/nlm:nlmuid-101488763-bk.

44. Moore to Jones, February 17, 1863, in "Biographical Sketch of Joseph Jones."

45. Jones, *Researches upon "Spurious Vaccination,"* 3.

46. "Richmond in Flames and Rubble," American Battlefield Trust, spring 2015, https://www.battlefields.org/learn/articles/richmond-flames-and-rubble. See also Paul D. Casdorph, *Confederate General R. S. Ewell: Robert E. Lee's Hesitant Commander* (Lexington: University Press of Kentucky, 2015), 331.

47. Jones, *Researches upon "Spurious Vaccination,"* 4.

48. For more on Confederate nationalism, including the role of women during and after the war, see Drew Faust, *The Creation of Confederate Nationalism: Ideology and Identity in the Civil War South* (Baton Rouge: Louisiana State University Press,

1989); Karen L. Cox, *Dixie's Daughters: The United Daughters of the Confederacy and the Preservation of Confederate Culture* (Gainesville: University Press of Florida, 2003).

49. Jones, *Researches upon "Spurious Vaccination,"* 26, 32, 90, 104; Elisha Harris, a Northern medical professional, included a section in his report on Confederate medicine, drawn from Jones's work, in "Vaccination in the Army,"154–160.

50. Jones, *Researches upon "Spurious Vaccination,"* 9, 12–13.

51. Jones, *Researches upon "Spurious Vaccination,"* 12.

52. "Trial of Henry Wirz. Letter of the Secretary of War Ad Interim, in Answer to a Resolution of the House of Representatives of April 16, 1866, Transmitting a Summary of the Trial of Henry Wirz," Executive Document no. 23, House of Representatives, 2nd Session, 40th Congress, 1868 (Washington, DC: Government Printing Office, 1868), 3–8, available at https://www.loc.gov/rr/frd/Military_Law/pdf/Wirz-trial.pdf.

53. John Fabian Witt argues that "nearly 1,000 individuals were charged with violating the laws of war during the course of the conflict." Witt, *Lincoln's Code: The Laws of War in American History* (New York: Free Press, 2012), 267. See also Nicholas R. Doman, review of *The Nuremberg Trials,* by August von Knieriem, *Columbia Law Review* 60, no. 3 (1960), 419.

54. See James O. Breeden, "Andersonville—A Southern Surgeon's Story," *Bulletin of the History of Medicine* 47, no. 4 (1973): 317–343.

55. "Trial of Henry Wirz," 4–5.

56. "Trial of Henry Wirz," 178.

57. "Trial of Henry Wirz," 663, 665, 667.

58. Jones, *"Researches upon "Spurious Vaccination,"* 14.

59. Jones, *"Researches upon "Spurious Vaccination,"* 15–16.

60. "Trial of Henry Wirz," 618.

61. "Trial of Henry Wirz," 642.

62. Jones, *"Researches upon "Spurious Vaccination,"* 16.

63. Jones, *"Researches upon "Spurious Vaccination,"* 17. Stephanie McCurry makes this argument in her book on the collapse of the Confederacy. McCurry, *Confederate Reckoning: Power and Politics in the Civil War South* (Cambridge: Harvard University Press, 2010).

64. Shawn Buhr, "To Inoculate or Not to Inoculate?: The Debate and the Smallpox Epidemic of 1721," *Constructing the Past* 1, no. 1 (2000): 61–66. Also see Michael Willrich, *Pox: An American History* (New York: Penguin Press, 2011), 37–39; James Colgrove, "Between Persuasion and Compulsion: Smallpox Control in Brooklyn and New York, 1894–1902," *Bulletin of the History of Medicine* 78, no. 1 (2004): 349–378.

65. The epidemic spread more viciously in the South than in the North. While the Union Army did vaccinate its soldiers, infected soldiers were often quarantined. When Union doctors engaged in vaccinations, it was often—as in the case of Ira Russell—in the South or Border states. Downs, *Sick from Freedom*, 98; Harris, "Vaccination in the Army."

66. Jones, *Researches upon "Spurious Vaccination,"* 13.

67. Jowan G. Penn-Barwell, "Sir Gilbert Blane FRS: The Man and His Legacy," *Journal of the Royal Naval Medical Service* 102, no. 1 (2016): 61–66.

68. Jones, *Researches upon "Spurious Vaccination,"* 17.

69. Jones, *Researches upon "Spurious Vaccination,"* 18–23.

70. Jones, *Researches upon "Spurious Vaccination,"* 24.

71. Jones, *Researches upon "Spurious Vaccination,"* 24.

72. Jones, *Researches upon "Spurious Vaccination,"* 25–27.

73. "Trial of Henry Wirz," 760–761.

74. "Trial of Henry Wirz," 775–777.

75. "Trial of Henry Wirz," 5.

76. In *Researches upon "Spurious Vaccination,"* Jones refers to harvesting lymph from infants on pages 27 ("lymph from a healthy infant" collected during the war), 76 (report by a Dr. Bigelow of a five-month-old "healthy child" who died of erysipelas after removal of vaccine matter, and by a Dr. Homans of a three-week-old infant who was vaccinated, had lymph matter removed, and then became ill with erysipelas), and 86 (obtaining lymph "from the arms of healthy children and infants" during the war).

77. Jones, *Researches upon "Spurious Vaccination."*

8. Narrative Maps

1. On Confederate wives, see Caroline E. Janey, *Remembering the Civil War: Reunion and the Limits of Reconciliation* (Chapel Hill: University of North Carolina Press, 2016), 92. On pension applications, see Theda Skocpol, *Protecting Soldiers and Mothers: The Political Origins of Social Policy in the United States,* rev ed. (Cambridge: Harvard University Press, 1995), 108–116. On Black veterans, see Jim Downs, *Sick from Freedom: African American Illness and Suffering during the Civil War and Reconstruction* (New York: Oxford University Press, 2012), 155.

2. Edmund Charles Wendt, ed., *A Treatise on Asiatic Cholera* (New York: William Wood, 1885), iv.

3. Wendt, *Treatise on Asiatic Cholera,* iii, v.

4. Wendt, *Treatise on Asiatic Cholera,* iv–v.

5. Ely McClellan, "A History of Epidemic Cholera, as It Affected the Army of the United States," part 1, section 2 in Wendt, *Treatise on Asiatic Cholera,* 71.

6. McClellan, "History of Epidemic Cholera," 71–78.

7. John W. Hall, *Uncommon Defense: Indian Allies in the Black Hawk War* (Cambridge, MA: Harvard University Press, 2009).

8. McClellan, "History of Epidemic Cholera," 78. Other accounts verify the presence of cholera in Indian Country and confirm that military troops transmitted the disease. See John C. Peters, "Conveyance of Cholera from Ireland to Canada and the United States Indian Territory, in 1832," *Leavenworth Medical Herald,* October 1867, 3. On Jackson's policies in the context of cholera in this region, see Ann Durkin Keating and Kathleen A. Brosnan, "Cholera and the Evolution of Early Chicago," in *City of Lake and Prairie: Chicago's Environmental History*, ed. Kathleen A. Brosnan,

William C. Barnett, and Ann Durkin Keating (Pittsburgh: University of Pittsburgh Press, 2020), 26–29.

9. McClellan, "History of Epidemic Cholera," 101.

10. McClellan, "History of Epidemic Cholera," 101–108.

11. McClellan, "History of Epidemic Cholera," 104.

12. McClellan, "History of Epidemic Cholera," 101, 107.

13. On refugee camps for newly emancipated enslaved people, see Downs, *Sick from Freedom*. On formerly enslaved people within Native American communities, see Barbara Krauthamer, *Black Slaves, Indian Masters: Slavery, Emancipation, and Citizenship in the Native American South* (Chapel Hill: University of North Carolina Press, 2015); Tiya Miles and Sharon P. Holland, eds., *Crossing Waters, Crossing Worlds: The African Diaspora in Indian Country* (Durham, NC: Duke University Press, 2006).

14. For more on how the nineteenth century led to a dramatic shift in maritime traffic and trade, see Jürgen Osterhammel, *The Transformation of the World: A Global History of the Nineteenth Century* (Princeton, NJ: Princeton University Press, 2014).

15. Most historians chart the origin of medical surveillance as a phenomenon that began in the twentieth century with the rise of bacteriology. The outbreak of tuberculosis in the early twentieth century, according to some scholars, marked the beginning of the government's systematic surveillance efforts. See, for example, Amy L. Fairchild, Ronald Bayer, and James Colgrove, *Searching Eyes: Privacy, the State, and Disease Surveillance in America* (Berkeley: University of California Press, 2007), 33. The outbreak of cholera in 1866 proves this not to be true.

16. Gavin Milroy, "The International Quarantine Conference of Paris in 1851–52, with Remarks," *Transactions of the National Association for the Promotion of Social Science,* ed. George W. Hastings (London: John W. Parker and Son, 1860), 606.

17. Milroy, "International Quarantine Conference of Paris," 606.

18. Milroy, "International Quarantine Conference of Paris," 608.

19. Milroy, "International Quarantine Conference of Paris"; Valeska Huber, "The Unification of the Globe by Disease? The International Sanitary Conferences on Cholera, 1851–1894," *The Historical Journal* 49, no. 2 (2006), 460.

20. David Arnold, *Colonizing the Body: State Medicine and Epidemic Disease in Nineteenth-Century India* (Berkeley: University of California Press, 1993), 169–170; Gyan Prakash, *Another Reason: Science and the Imagination of Modern India* (Princeton, NJ: Princeton University Press, 1999), 137–138; Edward Said, *Orientalism* (New York: Vintage, 1979).

21. Huber, "Unification of the Globe by Disease?" 461.

22. Deborah Jenson, Victoria Szabo, and the Duke FHI Haiti Humanities Laboratory Student Research Team, "Cholera in Haiti and Other Caribbean Regions, 19th Century," *Emerging Infectious Diseases* 17, no. 11 (2011): 2130–2135; "Cholera in South America," *JAMA* 8, no. 6 (February 5, 1887): 155–156; Charles Rosenberg, *The Cholera Years: The United States in 1832, 1849, and 1866* (Chicago: University of Chicago Press, 1962).

23. Huber, "Unification of the Globe by Disease?" 462.

24. Huber, "Unification of the Globe by Disease?" 463. From the vantage point of Europe, Russia was becoming a watchdog for cholera, but by taking on this role, Russia also expanded its empire. See Eileen Kane, *Russian Hajj: Empire and the Pilgrimage to Mecca* (Ithaca, NY: Cornell University Press, 2015), 45–46.

25. Norman Howard-Jones, *The Scientific Background of the International Sanitary Conference, 1851–1938* (Geneva: WHO, 1975), 31–33.

26. J. Netten Radcliffe, "Report on the Recent Epidemic of Cholera (1865–1866)," read April 6, 1868, *Transactions of the Epidemiological Society of London* vol. 3: Sessions 1866 to 1876 (London: Hardwicke and Bogue, 1876), 236–237.

27. Radcliffe, "Report on the Recent Epidemic of Cholera," 232.

28. Radcliffe, "Report on the Recent Epidemic of Cholera," 232, quoting from a memo that Radcliffe and Gavin Milroy presented to the Foreign Office in 1865.

29. Radcliffe, "Report on the Recent Epidemic of Cholera," 234–235.

30. Radcliffe, "Report on the Recent Epidemic of Cholera," 237–238.

31. Radcliffe, "Report on the Recent Epidemic of Cholera," 238.

32. Radcliffe, "Report on the Recent Epidemic of Cholera," 239.

33. "H.M.S. Penguin in the Gulf of Aden," *Illustrated London News* 50, no. 1433 (June 29, 1867): 648–649,

34. Matthew S. Hopper, *Slaves of One Master: Globalization and Slavery in Arabia in the Age of Empire* (New Haven, CT: Yale University Press, 2015), 169.

35. Hopper, *Slaves of One Master;* Radcliffe, "Report on the Recent Epidemic of Cholera," 238–239.

36. I draw here on the concept of "framing disease." See Charles Rosenberg and Janet Golden, eds., *Framing Disease: Studies in Cultural History* (New Brunswick, NJ: Rutgers University Press, 1992).

37. Radcliffe, "Report on the Recent Epidemic of Cholera," 239.

38. Radcliffe, "Report on the Recent Epidemic of Cholera," 239.

39. Radcliffe, "Report on the Recent Epidemic of Cholera," 240.

40. Radcliffe, "Report on the Recent Epidemic of Cholera," 240–241.

41. Radcliffe, "Report on the Recent Epidemic of Cholera," 241–243.

42. Radcliffe, "Report on the Recent Epidemic of Cholera," 243.

43. Radcliffe, "Report on the Recent Epidemic of Cholera," 244.

44. Radcliffe, "Report on the Recent Epidemic of Cholera," 244.

45. For a definition of Takruri, see 'Umar Al-Naqar, "Takrūr: The History of a Name," *Journal of African History* 10, no. 3 (1969): 365–374.

46. Radcliffe, "Report on the Recent Epidemic of Cholera," 244; [Henry] Blanc, *The Story of the Captives: A Narrative of the Events of Mr. Rassam's Mission* (London: Longmans, Green, Reader, and Dyer, 1868). On Blanc, see E. R. Turton, review of *A Narrative of Captivity in Abyssinia* by Henry Blanc, *Transafrican Journal of History* 2, no. 1 (1972), 135–136.

47. Recent advances in phylogenetics have shown that the Black Death and other plague pandemics stretched across a much wider terrain and lasted for a longer period than had been thought. This illustrates how contemporary scholars have often pieced together past pandemics, unlike Radcliffe, who did it on his own based on reports generated by the British Empire. See Monica H. Green, "The Four Black

Deaths," *American Historical Review* 125, no 5 (2020): 1601–1631; William McNeill, *Plagues and Peoples* (New York: Anchor, 1976).

48. Radcliffe, "Report on the Recent Epidemic of Cholera," 245.

49. Judith Salerno and Paul Theerman, "Looking Out for the Health of the Nation: The History of the U.S. Surgeon General," Books, Health, and History, New York Academy of Medicine, October 23, 2018, https://nyamcenterforhistory.org/2018/10 /23/surgeon-general. While some have blamed the surgeon general for not responding effectively to the initial health crises during the war, the federal government had limited authority and power to intervene in the daily lives of ordinary citizens or even to establish a system of medical surveillance powered by a military bureaucracy. On criticism of the surgeon general, see Louis C. Duncan, "The Medical Department of the United States Army in the Civil War—The Battle of Bull Run," *The Military Surgeon* 30, no. 6 (1912): 644–668. See also Chapter 6 on the USSC.

50. For more on the Medical Division of the Freedmen's Bureau and its response to the smallpox epidemic, see Downs, *Sick from Freedom*.

51. Downs, *Sick from Freedom*, 95–119.

52. J. J. Woodward, *Report on Epidemic Cholera in the Army of the United States, during the Year 1866* (Washington, DC: Government Printing Office, 1867), iii, v (hereafter, Woodward, *Report on Cholera 1866*).

53. Elizabeth Fenn documents the spread of smallpox during the American War of Independence, but the military did not develop the medical surveillance technology, namely a highly advanced bureaucracy, that tracked the spread of smallpox across the new nation and rest of North America. More to the point, Fenn's study examines the years between 1775 and 1782 and Congress did not establish the Surgeon General's Office until 1798. See Elizabeth Fenn, *Pox Americana: The Great Smallpox Epidemic, 1775–1782* (New York: Hill and Wang, 2002).

54. The outbreak of yellow fever in 1793 reveals the limitations of city governments to respond to epidemics and to institute effective protocols within their own jurisdictions, let alone beyond their borders. See J. M. Powell, *Bring Out Your Dead: The Great Plague of Yellow Fever in Philadelphia in 1793* (1949; repr., Philadelphia: University of Pennsylvania Press, 1993), 55–63, 173–194. See also Rosenberg, *Cholera Years*, 82, 90.

55. Rosenberg, *Cholera Years*, 22.

56. Woodward, *Report on Cholera 1866*, v–vi.

57. Woodward, *Report on Cholera 1866*, vi, 29–35.

58. Woodward, *Report on Cholera 1866*, vii.

59. Woodward, *Report on Cholera 1866*, viii.

60. Woodward, *Report on Cholera 1866*, xvi.

61. Woodward, *Report on Cholera 1866*, ix.

62. Woodward, *Report on Cholera 1866*, ix, xii.

63. Woodward, *Report on Cholera 1866*, viii.

64. James G. Hollandsworth Jr., *An Absolute Massacre: The New Orleans Race Riot of July 30, 1866* (Baton Rouge: Louisiana State University Press, 2001), 123, 129.

65. Woodward, *Report on Cholera 1866*, vii–viii.

66. Woodward, *Report on Cholera 1866*, xi, 49.

67. Woodward, *Report on Cholera 1866*, ix.

68. Woodward, *Report on Cholera 1866*, xi–xiii, 59–60.

69. See, for example, Woodward, *Report on Cholera 1866*, 32, 33.

70. Woodward, *Report on Cholera 1866*, 30.

71. For references to rice-water discharge as a distinguishing characteristic of cholera, see Woodward, *Report on Cholera 1866*, 21, 23, 26, 53.

72. Woodward, *Report on Cholera 1866*, 25.

73. Woodward, *Report on Cholera 1866*, 57.

74. Woodward, *Report on Cholera 1866*, 39–40 (water in New Orleans), 59 (water in Nicaragua), xvii (Woodward quotes).

75. "Dr. Benjamin Faneuil Craig," *Boston Medical and Surgical Journal* 96 (May 17, 1877): 590–592.

76. Woodward, *Report on Cholera 1866*, xvii, 62.

77. Woodward, *Report on Cholera 1866*, xvii.

78. On the problems of distributing supplies throughout the postwar South, see Downs, *Sick from Freedom*, 65–94.

79. Woodward, *Report on Cholera 1866*, 30.

80. Woodward, *Report on Cholera 1866*, 35.

81. J. J. Woodward, *Report on Epidemic Cholera and Yellow Fever in the Army of the United States, during the Year 1867* (Washington, DC, Government Printing Office, 1868), 23, 38, 50.

82. Woodward, *Report on Cholera 1866*, xiii; Woodward, *Report on Epidemic Cholera and Yellow Fever . . . during the Year 1867*, vi.

83. Woodward, *Report on Epidemic Cholera and Yellow Fever . . . during the Year 1867*, vi.

84. An exception is Dr. William B. Fletcher, who published a study of nineteenth-century cholera epidemics in 1866 that includes a foldout map tracing the course of the four epidemics from east to west, but he notes that "This map does not represent each city or each small division of a country that has suffered. . . . The intention is merely to give a general idea of the main current of the epidemic." William B. Fletcher, *Cholera: Its Characteristics, History, Treatment, Geographical Distribution of Different Epidemics, Suitable Sanitary Preventions, etc.* (Cincinnati: Robt. Clarke & Co, 1866).

85. Steven Johnson, *The Ghost Map: The Story of London's Most Terrifying Epidemic—and How It Changed Science, Cities, and the Modern World* (New York: Riverhead, 2006). There is some debate about the efficacy of Snow's maps; see Tom Koch and Ken Denike, "Essential, Illustrative, or . . . Just Propaganda? Rethinking John Snow's Broad Street Map," *Cartographica* 45, no. 1 (2010): 19–31.

Conclusion

1. See, for example, Warwick Anderson, *Colonial Pathologies: American Tropical Medicine, Race, and Hygiene in the Philippines* (Durham, NC: Duke University Press, 2006); Jennifer L. Morgan, *Laboring Women: Reproduction and Gender in New World Slavery* (Philadelphia: University of Pennsylvania Press, 2004); Rana A. Hogarth,

Medicalizing Blackness: Making Racial Difference in the Atlantic World (Chapel Hill: University of North Carolina Press, 2017).

2. See, for example, Sven Beckert and Seth Rockman, *Slavery's Capitalism: A New History of American Economic Development* (Philadelphia: University of Pennsylvania Press, 2016); Edward E. Baptist, *The Half Has Never Been Told: Slavery and the Making of American Capitalism* (New York: Basic Books, 2014); Daina Ramey Berry, *The Price for Their Pound of Flesh: The Value of the Enslaved, from Womb to Grave, in the Building of a Nation* (Boston: Beacon Press, 2017); Eric Williams, *Capitalism and Slavery* (Chapel Hill: University of North Carolina Press, 1994).

3. Beckert and Rockman, *Slavery's Capitalism*, 3.

4. Craig Steven Wilder, *Ebony and Ivy: Race, Slavery, and the Troubled History of America's Universities* (New York: Bloomsbury Press, 2014); Nancy Stepan, *The Hour of Eugenics: Race, Gender, and Nation in Latin America* (Ithaca, NY: Cornell University Press, 1996).

5. Deirdre Cooper Owens, *Medical Bondage: Race, Gender, and the Origins of American Gynecology* (Athens: University of Georgia Press, 2017); Harriet Washington, *Medical Apartheid: The Dark History of Medical Experimentation on Black Americans from Colonial Times to the Present* (New York: Anchor, 2008); Ann Fabian, *The Skull Collectors: Race, Science, and America's Unburied Dead* (Chicago: University of Chicago Press, 2010); Hogarth, *Medicalizing Blackness*.

6. Dorothy Roberts, *Fatal Invention: How Science, Politics, and Big Business Re-create Race in the Twenty-first Century* (New York: New Press, 2011).

7. John Duffy, *The Sanitarians: A History of American Public Health* (Urbana: University of Illinois Press, 1990); George Rosen, *A History of Public Health*, rev. expanded ed. (Baltimore: Johns Hopkins University Press, 2015).

8. On the use of racial ideology as an explanatory force, see Karen E. Fields and Barbara Jeanne Fields, *Racecraft: The Soul of Inequality in American Life* (New York: Verso, 2012).

9. These practices originated because of violence enacted both by military powers and by what Gayatri Spivak refers to as "epistemic violence": a violation in the process of knowledge production, an "asymmetrical obliteration of the trace of that Other in its precarious Subject-ivitiy." Gayatri Chakravorty Spivak, "Can the Subaltern Speak?" in *Marxism and the Interpretation of Culture,* ed. Cary Nelson and Lawrence Grossberg (Urbana: University of Illinois Press, 1988), 271–313.

Acknowledgments

This book began when I came across reports of the 1865–1866 cholera pandemic while completing my first book, *Sick from Freedom: African American Illness and Suffering during the Civil War and Reconstruction*. I became increasingly interested in the origins of epidemiology and the ways British physicians developed effective protocols that US doctors later adopted. I wanted to determine if there had been any discussion of race, which was often buried in the US archives. Lesley Hall, a wise archivist at the Wellcome Institute for the History of Medicine in London, directed me to the Medical Officer of Health Reports in the National Archives at Kew, England. From my first visit in 2011 to my final visit in 2019, the archivists and librarians were superb. My first debt is to the archivists at the Wellcome, the National Archives, and the British Library, and to their meticulous preservation and organization of records. I am also very grateful to Mary Holt, the history librarian at Tulane University's Rudolph Matas Library of the Health Sciences, who spent days helping me sift through the collections, and to Arlene Shaner, historical collections reference librarian at the New York Academy of Medicine for her enthusiasm and support.

The Andrew W. Mellon New Directions Fellowship provided me with a magnanimous fellowship that enabled me to return to graduate school, after tenure, to gain postgraduate training in medical anthropology at Harvard University, where I was also a visiting fellow at the Weatherhead Initiative on Global History (WIGH). I took classes with epidemiologists, medical anthropologists, physicians, and scholars in medical humanities who expanded my knowledge of medicine and health. I am most grateful to Arthur Kleinman, Paul Farmer, David Jones, and most especially Allan M. Brandt and Evelynn Hammonds. I delivered the first iteration of the main argument for this book while I was a fellow at WIGH. Evelynn Hammonds and Westenley Alcenat provided insightful comments that generated a productive conversation. That presentation would not have been possible without Sven Beckert, who invited me to be a fellow at WIGH, encouraged my journey into global history, and raised smart questions that have shaped the direction of this book. Elizabeth Hinton was my main interlocutor at Harvard and a dear friend, who helped me think through the argument of this book. After I left Harvard, Evelyn Brooks Higginbotham has continued to support me and this project; without her help, I am convinced this book would not have been completed. Her intellectual generosity and brilliant scholarship have made an indelible imprint on this project.

In 2012, I was a visiting fellow in the Department of the History of Science at Harvard University. Charles Rosenberg graciously met with me and helped me evaluate some sources on cholera; he also provided brilliant comments during the talk I presented in the department's Incubator Series, wonderfully organized by Sophia Roosth.

I have presented portions of the material in this book at a number of scholarly meetings. I am grateful for the enthusiastic responses from colleagues and friends at the American Association for the History of Medicine, American Historical Association, Organization of American Historians, American Studies Association, C19, and Society for Historians of the Early American Republic, as well as the "Medicine and Healing in the Age of Slavery" conference at Rice University in 2018, the Early Atlantic Seminar at the University of Michigan, the 2019 Draper Conference on "Greater Reconstruction: American Democracy after the Civil War" held at the University of Connecticut, and the

"Closer to Freedom" conference held in honor of the late Stephanie M. H. Camp at the University of Pennsylvania in 2015. In particular, I would like to thank Rana Hogarth, Christopher Willoughby, Urmi
Engineer Willoughby, Leslie Harris, Stacey Patton, Sowande' Mustakeem, Carla Peterson, Julia Rosenbloom, Shauna Devine, Todd L.
Savitt, Sharla Fett, Gretchen Long, Savannah Williamson, Sean Morey
Smith, Stephen Kenny, Jon Wells, Adam Biggs, Crystal Feimster,
Jessica Marie Johnson, Manisha Sinha, and Vanessa Gamble. A special
acknowledgment to the brilliant Keith Wailoo; when I was working on
my first book, he provided a formal comment at the American Historical
Association meeting that provided an extremely useful framework for interpreting sources. Similarly, when I began writing this manuscript,
he provided a formal comment at the American Studies Association
meeting in 2015 that profoundly shaped how I framed this book.

Yael Sternhell invited me to Tel Aviv University, where I was able to
give a presentation to a group of engaged researchers and smart students. I am also grateful to the Association of British American Nineteenth Century Historians for inviting me to deliver the 2018 Parish Lecture at Cambridge University, which provided me with an opportunity
to share my research with colleagues across the pond. Edward Rugemer
invited me to present my research at the "Race and Slavery in the Atlantic World to 1900" workshop at Yale, where an enthusiastic audience
asked probing questions and offered encouraging feedback.

Since my first book, Margaret Humphreys has always been willing
to answer questions or think through an idea with me. For this project,
she kindly shared with me her typed transcriptions of the US Sanitary
Commission Papers. Robert D. Hicks, former director of the Mütter
Museum and the Wood Institute for the History of Medicine at the College of Physicians of Philadelphia, has spent many hours talking to me
about smallpox vaccinations and helped me locate a source in the middle
of the pandemic.

After writing Chapter 1 in isolation for almost a year, presenting my
research at the "Trans-America Crossings" conference organized by the
Omohundro Institute and the John Carter Brown Library, Brown University, in June 2018 was a critical turning point. I am deeply grateful to
my fellow panelists, Carolyn Roberts and Marisa Fuentes, whose enthusi-

astic reaction to my presentation inspired me enormously. Jennifer L. Morgan chaired the panel; her research has been a model for me, and her prodigious contributions to the historiography of slavery have taught me a great deal. Her support has emboldened me to do my best.

From the first time I delivered a version of Chapter 3 at the "Closer to Freedom" conference in honor of the late Stephanie Camp in 2015, Kathleen M. Brown has enthusiastically supported this project. She has encouraged me to think globally while keeping my focus on the everyday actions of ordinary people. A conversation we once had about washerwomen's work has sustained a key argument in this book. As always, I am deeply grateful for her friendship and intellectual camaraderie.

My research for this book unfolded at the same time that Deirdre Cooper Owens was finishing *Medical Bondage,* her award-winning study of slavery and gynecology. We have traveled together from Belfast to Los Angeles to Philadelphia to Raleigh, always discussing the history of race and medicine. Her contributions to this book can be found not only in the endnotes but also indelibly marked on the pages.

Ever since I delivered my first paper on this project at the American Association for the History of Medicine conference in 2015, Jacob Steere-Williams and I have been in conversation about the history of epidemiology. He has provided encouragement and helped me interrogate the sources and build an analytical framework. This book is much better because of his profound influence. I am lucky not only to have benefited from his capacious intellect and analytical rigor but also to have found a friend in the process.

Breakfast meetings at conferences and long phone calls with Susan O'Donovan have been instrumental for writing this book; her inimitable energy and impressive analytical mind have been a constant source of nourishment. Many phone conversations with Monique Bedasse as well as impromptu meetings in London and New York about the Caribbean have been enormously instructive. Her friendship and immeasurable support are a blessing. Seth Rockman has discussed my overall argument with me on many occasions, suggested articles and books to read, and carefully read key parts of the manuscript. While I remain grateful for his friendship, his brilliant mind has helped me to articulate critical passages of the book.

At Connecticut College, I was fortunate to have colleagues who have supported me over the last decade. I am most grateful to Sunil Bhatia, Sandy Grande, David Kim, Aida Heredia, Afshan Jafar, Denise Pelletier, Michelle Neely, Cherise Harris, Nancy Lewandowski, Rosa Woodhams, and Cathy Stock. Marc Forster would often generously stop what he was doing to teach me about European history. I am also particularly grateful to the office of Dean of the Faculty for support over the years, particularly Deans Jeff Cole and Anne Bernhard. A special note of appreciation to Katherine Bergeron, whose interest in my research and writing has made all the difference.

Through the generous support of a ConnSSHARP grant, I was able to hire Carter Goffigon, class of 2014, who spent the summer of 2013 at the New York Academy of Medicine poring over early nineteenth-century medical treatises about cholera and infectious disease with me. Her intellectual sophistication continues to impress me. In 2018, ConnSSHARPs and CELS grants allowed me to work with three superb undergraduate research assistants—Jackson Bistrong, Miles Hamberg, and Max Amar-Olkus—who helped me examine the US Sanitary Commission Papers at the New York Public Library. Jennifer Nichole provided helpful, enthusiastic research assistance at the Archivo General de Indias in Seville. Thanks also to the National Endowment for the Humanities Summer Fellowship, and the R. F. Johnson and Research Matters awards from Connecticut College for funding this project.

I finished this book as the Gilder Lehrman-NEH Professor of Civil War Studies and History at Gettysburg College. Peter Carmichael has been my main intellectual comrade and friend; his enthusiasm and support mean the world to me. I am also grateful to Kari Greenwalt, Scott Hancock, Michael Birkner, Heather Miller, Chris Zappe, and the history department and staff at the Civil War Institute. My students in my fall 2020 "Narratives of Illness" class were a collection of some of the sharpest, most dazzling minds whom I have ever taught. Their engagement with the readings in the middle of a pandemic made every class a thrill to teach—thank you!

At Harvard University Press, I am deeply grateful to my editor, Janice Audet, for her unfailing ebullience, editorial rigor, and intellectual generosity. Emeralde Jensen-Roberts has been an excellent asset to the production of this book. Louise E. Robbins is a brilliant manuscript editor. Working with her has been one of the richest, most intellectually rewarding endeavors of my career. My praise of her editorial sophistication cannot begin to convey my deep gratitude for her careful examination of the manuscript. Thomas LeBien was my editor at HUP before his departure. He was an early champion of the book and immediately understood the argument. His infectious curiosity, sharp analytical mind, and capacious historical imagination have guided this project.

Anne Kornhauser has been discussing this book with me since its inception, offering ingenious, judicious, and encouraging feedback. Monica R. Gisolfi has insightfully helped me think through the chronology of this story and its relationship to larger historiographical polemics. The late Thea K. Hunter was always bursting with boundless questions and comments whenever we talked about this project. Her unexpected death created an irrevocable deficit in my research and life.

Over the years, many colleagues and mentors have supported my research. I am deeply indebted to Eric Foner, Erica Armstrong Dunbar, Thavolia Glymph, David Blight, Allyson Hobbs, Megan Kate Nelson, Lázaro Lima, and Heather Ann Thompson. I am also deeply indebted to my friends and family, who have cheered me on while I wrote—most particularly Andrea Dalimonte, Brandon LoCasto, Joe Figini, John Bantivoglio, Liana Dao, George Davilas, Ernie Alverez, Jorge Campos, Todd Anten, Sally Anne French, Jaclynn Downs, Jeff Downs, Jada Downs, my parents, and my extended family.

Catherine Clinton has been a wonderful, supportive mentor. She has served as a sounding board, an editor, a colleague, a teacher, and, most of all, a friend. My research and writing are stronger because of her dazzling influence and brilliant mind. She once asked me, "What about Antoine Lavoisier?" The intellectual journey I embarked on to answer that question resulted in Chapter 1 of this book.

While I was writing this manuscript, both of my parents were diagnosed with cancer within months of each other. Since 2017, I have spent a great deal of time in hospitals. If I were a poet, I would be able to develop a metaphor to connect their health to the people, diseases, and medical practices in this book, but I can only organize the past—not place it in a lyrical dialogue with the present. Trying to juggle writing with spending time in hospital rooms, waiting rooms, and hospital cafeterias was difficult. But I found the endurance to press ahead by remembering the struggles of my ancestors. They have been with me while I was alone in a hospital café or in an archive or in a lonely city, their own lives serving as an example for me to follow. Each time I boarded a flight to London, crossed the dark Atlantic Ocean, and arrived at the archives, I realized the documents I was planning to read were not intended for me. I have dedicated this book to my ancestors because my decision to become a historian was not my choice alone. It was whispered to me. They have guided me through the archives and reminded me that writing a book is a privilege only made possible as a direct result of their struggles and gifts.

Index